Ecology

Laboratory Manual

Ecology

Laboratory Manual

Darrell S. Vodopich

Baylor University

Boston Burr Ridge, IL Dubuque, IA New York San Francisco St. Louis
Bangkok Bogotá Caracas Kuala Lumpur Lisbon London Madrid Mexico City
Milan Montreal New Delhi Santiago Seoul Singapore Sydney Taipei Toronto

ECOLOGY LABORATORY MANUAL

Published by McGraw-Hill, a business unit of The McGraw-Hill Companies, Inc., 1221 Avenue of the Americas, New York, NY 10020.

Some ancillaries, including electronic and print components, may not be available to customers outside the United States.

 This book is printed on recycled, acid-free paper containing 10% postconsumer waste.

1 2 3 4 5 6 7 8 9 0 QPD/QPD 0 9

ISBN 978–0–07–338318–7
MHID 0–07–338318–X

Publisher: Janice Roerig-Blong
Executive Editor: Margaret J. Kemp
Director of Development: Kristine Tibbetts
Developmental Editor: Fran Schreiber
Marketing Manager: Heather Chase Wagner
Project Manager: Joyce Watters
Senior Production Supervisor: Kara Kudronowicz
Associate Design Coordinator: Brenda A. Rolwes
Cover Designer: Studio Montage, St. Louis, Missouri
(USE) Cover Image: © Digital Vision/Getty Images
Senior Photo Research Coordinator: Lori Hancock
Compositor: Lachina Publishing Services
Typeface: 10/12 Times Roman
Printer: Quebecor World Dubuque, IA

Some of the laboratory experiments included in this text may be hazardous if materials are handled improperly or if procedures are conducted incorrectly. Safety precautions are necessary when you are working with chemicals, glass test tubes, hot water baths, sharp instruments, and the like, or for any procedures that generally require caution. Your school may have set regulations regarding safety procedures that your instructor will explain to you. Should you have any problems with materials or procedures, please ask your instructor for help.

www.mhhe.com

Contents

Preface

I designed this manual to survey basic field and laboratory techniques for an introductory ecology course. The experiments and procedures are safe, easy to perform, and target the needs of undergraduate classes. The manual includes photographs, traditional topics, and a few broad-based exercises targeted for the wide context of life science. Each exercise has multiple, discrete procedures that help instructors tailor the exercise to students' needs, the style of the instructor, and the time and facilities available.

TO THE STUDENT

This manual introduces you to a part of life science that you probably don't know much about. As you learn about ecology, you'll spend equal time observing plants and animals around you and figuring out how to quantify their distribution, abundance, and interactions. Don't hesitate to exceed the observations outlined in the procedures—your future as a scientist depends on noticing things that others may overlook. In other words, don't underestimate the role of simple observation to support well-designed data collection and analysis. Now is the time to sharpen your skills in science with a mix of work and relaxed observation. Have fun, and learning will come easily. Also, remember that this manual is designed with both instructors and students in mind. Go to your instructors often with questions—their experience is a valuable tool that you should use as you work.

TO THE INSTRUCTOR

It has always bothered me that available ecology lab manuals are overwhelming in presentation. They are either too customized to use in different environments, too mathematical to let the biology of organisms and their environment shine through, or too vague to clearly lay out the steps of fundamental field and lab procedures we often use to answer ecological questions. So, as many of us have done, I designed my own. I hope you find it useful.

This manual is all about observing the natural world, asking questions, learning quantitative field techniques, and melding these activities into good science. The following design features of each exercise support this goal:

- First and foremost, instructors at any school with access to a stream, a pond, some woodlands, and some grasslands can do these exercises. Specialized habitats are not needed and equipment is minimal.
- This is an introductory manual. It assumes no previous knowledge of ecology, but it works best if the students have completed introductory biology.
- The procedures work. The techniques and instructions in the manual and the accompanying resource guide are detailed and straightforward, enough so that a knowledgeable teaching assistant can set up and supervise the techniques. However, this is not an autotutorial manual. Students are often directed to speak with their instructor. No manual can replace a real, live instructor.
- A few of the exercises are purposely designed for sessions in the lab without field work. Real-world teaching situations occasionally call for indoor sessions during the semester.
- Most exercises include enough variety of procedures to allow the instructor to pick and choose. The procedures are stand-alone, and customizing within or between exercises is rather easy.
- Each exercise involves a clear and rather singular concept. The introduction highlights that concept, and does it concisely. One exercise cannot completely survey a broad topic such as gas exchange, competition, or population growth, but I hope they clearly compliment the students' accompanying lecture course.
- The math required of the student does not overwhelm observations of ecology and adaptations of real organisms. Ecology is certainly a quantitative science, and tables with calculations abound throughout the manual. But a purposeful balance is struck between the need for quantification and trying to avoid massive calculations and statistics that mask learning of fundamental data-collection techniques.
- Questions are included with each procedure to bring students into the interpretive stage of ecological studies. Be sure to examine the "Questions for Further Thought and Study" in each exercise. They can expand students' perceptions that each exercise has broad applications to their world.

In summary, this manual's straightforward approach emphasizes experiments and activities that optimize students' investment of time and your investment of supplies, equipment, and preparation. Remind your teaching assistants that discussions and interactions between student and instructor are major components of a successful laboratory experience.

REVIEWERS

We thank the following reviewers for their helpful comments and suggestions during the preparation of this laboratory manual.

David L. Boose, Gonzaga University
Scott Burt, Truman State University
James R. Curry, Franklin College
Elizabeth L. Rich, Drexel University

Darrell S. Vodopich

Welcome to Field Work and the Ecology Laboratory

Welcome to ecology. Reading your textbook and attending lectures are certainly important ways to learn about ecology, but nothing can replace the importance of the laboratory. In lab you will get hands-on experience with what you've heard and read about ecology—for example, you'll observe organisms, learn techniques, test ideas, collect data, and draw conclusions about what you've learned. You'll do ecology.

You will enjoy the exercises in this manual—they're interesting, informative, and best of all they will introduce you to field work. I've provided questions to test your understanding of what you've done. In some of the exercises, I have also asked you to devise your own experiments to answer questions that you have posed. To make these exercises most useful and enjoyable, follow these guidelines:

THE IMPORTANCE OF COMING TO CLASS

The procedures in this manual are designed to help you experience ecology firsthand. To do that, you must attend class. If you want to do well in your ecology course, you'll need to attend class and pay attention. To appreciate the importance of class attendance for making a good grade in your ecology course, examine figure 1, a graph showing how students' grades in an introductory biology course relate to their rates of class attendance. Data are from a general biology class, University of Minnesota, 2003.

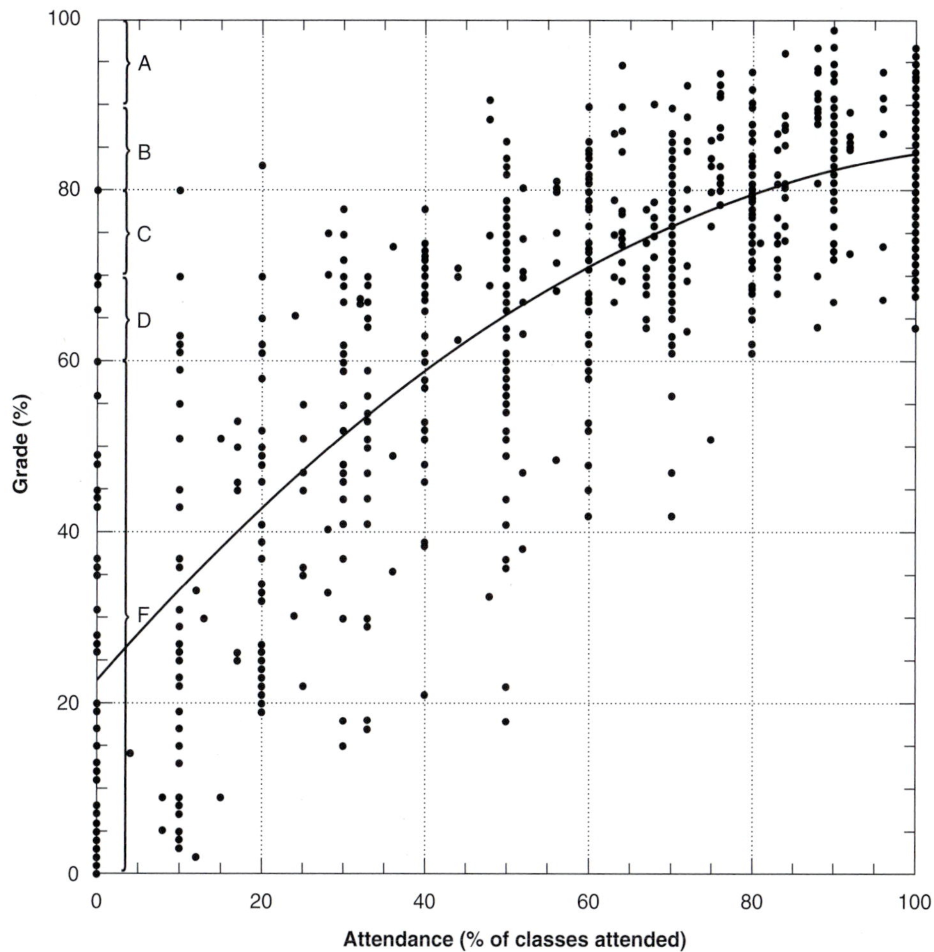

BEFORE COMING TO LAB

Read the exercise before coming to lab. This will give you a general idea about what you are going to do, and why you're going to do it. Knowing this will not only save time, it will also help you finish your work.

WHEN IN LAB

1. Don't start the exercise until you've discussed the exercise with your laboratory instructor. She/he will give you specific instructions about the work ahead and tell you if the exercise will be modified.
2. Work carefully and thoughtfully, and stay focused as you work. You can get your work done within the allotted time if you are well prepared and stay busy.
3. Discuss your observations, results, and conclusions with your instructor and lab partners. Perhaps their comments and ideas will help you better understand what you've observed.
4. Always follow instructions and safety guidelines presented by your instructor.
5. If you have questions, ask your instructor.

SAFETY IN THE LABORATORY

The laboratory safety rules listed in table 1 will help make lab a safe place for everyone to learn ecology. Remember, it is much easier to prevent an accident than to deal with its consequences.

Read the laboratory safety rules listed in table 1. If you do not understand them, or if you have questions, ask your instructor for an explanation. Then complete table 1 and sign the statement that is at the bottom of page xiii.

AFTER EACH LABORATORY

Soon after each lab, review what you did. What questions did you answer? What data did you gather? What conclusions did you make?

Also note any questions that remain. Try to answer these questions by using your textbook or by visiting the library. If you can't answer the questions, discuss them with your instructor.

Welcome to ecology!

Table 1

Laboratory Safety Rules

Rule	Why is this rule important? What could happen if this rule is not followed?
Behave responsibly. No horseplay or fooling around while in lab.	
Do not bring any food or beverages into lab, and do not eat, drink, smoke, chew gum, chew tobacco, or apply cosmetics when in lab. Never taste anything in lab. Do not put anything in lab into your mouth. Avoid touching your face, chewing on pens, and other similar behaviors while in lab.	
Unless you are told otherwise by your instructor, assume that all chemicals and solutions in lab are poisonous and act accordingly. Never pipette by mouth. always use a mechanical pipetting device (e.g., a suction bulb) to pipette solutions. Clean up all spills immediately, and report all spills to your instructor.	
Read the labels on bottles and know the chemical you are dealing with. Do not use chemicals from an unlabeled container, and do not return excess chemicals back to their container.	
Unless your instructor tells you to do otherwise, do not pour any solutions down the drain. Dispose of all materials as per instructions from your instructor.	
If you have long hair, tie it back. If you are using open flames, roll up loose sleeves. Wear contact lenses at your own risk; contacts hold substances against the eye and make it difficult to wash your eyes thoroughly.	
Treat living organisms with care and respect.	
Your instructor will tell you the locations of lab safety equipment, including fire extinguishers, fire blanket, eyewash stations, and emergency showers. Familiarize yourself with the location and operation of this equipment.	
Notify your instructor of any allergies to latex, chemicals, stings, or other substances.	
If you break any glassware, do not pick up the pieces of broken glass with your hands. Instead, use a broom and dustpan to gather the broken glass. Ask your instructor how to dispose of the glass.	
Unless told by your instructor to do otherwise, work only during regular, assigned hours when the instructor is present. Do not conduct any unauthorized experiments; for example, do not mix any chemicals without your instructor's approval.	
Do not leave any experiments unattended unless you are authorized to do so. If you leave your work area, slide your chair under the lab table. Keep walkways and desktops clean and clear by putting books, backpacks, and so on along the edge of the room, in the hall, in a locker, or in an adjacent room.	
Know how to use the equipment in lab. Most of the equipment is expensive; you may be required to pay all or part of its replacement cost. Keep water and solutions away from equipment and electrical outlets. Report malfunctioning equipment to your instructor. Leave equipment in the same place and condition that you found it. If you have any questions about or problems with equipment, contact your instructor.	
Know what to do and whom to contact if there is an emergency. Know the fastest way to get out of the lab. Immediately report all injuries—no matter how minor—to your instructor. Seek medical attention immediately if needed. If any injury appears to be life-threatening, call 911 immediately.	
At the end of each lab, clean your work area, wash your hands thoroughly with soap, slide your chair under the lab table, and return all equipment and supplies to their orignal locations. Do not remove any chemicals or equipment from the lab.	

Name ______________________________

Lab Section ______________________________

Your lab instructor may require that you submit this page at the end of today's lab.

1. In the space below, write an analysis of the data shown in figure 1 (page ix).

After completing table 1, sign this statement:

2. I have read, understood, and agree to abide by the laboratory safety rules described in this exercise and discussed by my instructor. I know the locations of the safety equipment and materials. If I violate any of the laboratory safety rules, my instructor may remove me from the lab.

____
Signature

____
Name (printed)

____
Date

W–5

The Nature of Data

1

Objectives

As you complete this lab exercise you will:

1. Calculate measures of central tendency for a data set and the different perspectives they provide on the same data set.
2. Understand the relationship between the mean and the variation for replicate values of a data set.
3. Examine the frequency distribution for three data sets.
4. Calculate the confidence intervals surrounding the mean of replicate values.

Ecologists collect data . . . and a lot of it. The data are extensive because (1) natural processes are complex; (2) ecological processes involve many variables; and (3) each variable can vary greatly. For example, plant growth is a complex process, the number of factors influencing growth is immense, and factors such as rainfall vary from day to day. To understand this kind of complexity and variation, ecologists analyze data to search for patterns and relationships among variables. In other words, they look at the nature of their data.

POPULATIONS AND SAMPLES OF POPULATIONS

Researchers gather data to describe and learn about large **populations**. Unfortunately, most populations are too big for us to measure a variable for every member of the population. For example, oak trees are too numerous for us to measure the length of *all* of their leaves. Instead, we **sample** the population of oak leaves. Sampling means that we take a relatively small number of measurements that represent the entire population. The characteristics we measure, such as leaf length, are **variables**, and the values for a variable are our **data**. Ecologists use these data to calculate **statistics** such as means, variances, etc. that describe our sample.

Statistics, such as a mean derived from samples, estimate variables of the entire population. For example, we measure the lengths of 20 sampled oak leaves and calculate a sample mean. This mean estimates the mean length of all the leaves in the population of oaks. We will never know the exact mean for the entire population so we use the sample mean ($\bar{x}$) to estimate the population mean (μ). Sample statistics are estimates of population statistics.

The major terms to quantitatively describe a set of data are:

population—the entire collection of possible measurements about which we wish to draw conclusions. All frogs in a pond, or all possible measurements of a forest's soil nitrate content, are examples of populations.

sample—a measurement(s) representing all possible measurements of a parameter of a population. It is a subset of all possible measurements in a population.

variable—a measurable characteristic of a biological entity. It may vary from one organism to the next, one environment to the next, or one moment to the next.

statistic—an estimate of a parameter based on a representative sample of a population. The mean of a set of values is a statistic.

DATA SETS

Data are usually organized into a **data set**, defined as a series of repeated measures of one or more variables. A variable might be the number of eggs in a robin's clutch, the concentration of nitrate in the soil, or the monthly rainfall on a prairie. You will quickly learn that the most noticeable attribute of a data set is its variation.

Procedure 1.1

Examine variation within a data set.

1. Examine the data set in table 1.1.
2. These data are measurements of the length of 20 leaves randomly selected from an oak tree. Notice that all leaves are not the same length.

3. Use the blanks provided in table 1.1 to rearrange and record the data from lowest to highest value.

Questions 1

Do the leaf lengths shown in table 1.1 appear to be simply random numbers? If not, what pattern or tendency do you detect? ______________________

____

What factors might cause variation in leaf length for an oak tree? ______________________

____

How would you sample leaves to test one of those influences on leaf length? ______________________

____

TABLE 1.1

A SAMPLE DATA SET OF 20 REPLICATE MEASUREMENTS OF OAK-LEAF LENGTHS

Oak Leaf Lengths:

78	69	62	74	69	51	45	40	9	64
___	___	___	___	___	___	___	___	___	___
65	64	61	69	52	60	66	71	72	27
___	___	___	___	___	___	___	___	___	___

Measures of Central Tendency:

Mean = ________ Median = ________ Mode = ________

MEASURES OF CENTRAL TENDENCY

The most likely "pattern" revealed by examining the data set in table 1.1 is the **central tendency** of the values. They are not spread out randomly. They tend to be clustered around a central value somewhere in the 50s, 60s, or 70s rather than randomly scattered from 1 to 100. That shouldn't surprise you—oak leaves don't grow randomly. Their development has a pattern.

The three most common measures of central tendency are mean, median, and mode. The **mean** ($\bar{x}$) is the arithmetic average of a group of measurements. It is the sum of all the values ($\sum x_i$) divided by the number of values (N).

$$\text{mean} = \bar{x} = \sum x_i / N$$

The **median** is the middle value of a group of measurements that have been ranked from lowest to highest or highest to lowest. The **mode** is the value that appears most often in the data set.

The mean is the most common measure of central tendency, but the median and mode are sometimes useful because they are less sensitive to extreme values. To appreciate the differences among the measures of central tendency, complete Procedure 1.2.

Procedure 1.2

Examine measures of central tendency of an example data set.

1. Calculate and record the mean of the data set in table 1.1.

Questions 2

Are any of the leaf measurements the same as the mean?

____

How many leaves were longer than the mean? ________

____

How many leaves were shorter than the mean? ________

____

Does the mean always describe the "typical" measurement? Why or why not? ______________________

____

____

2. Determine and record the median and mode of the data set in table 1.1.

Questions 3

Notice that in the sample, the mean differs from the median. What is responsible for this difference between the mean and the median? ______________________

____

How would the mean change without the 9-mm leaf?

____

____

How would the median change without the 9-mm leaf?

____

____

How would the mode change without the 9-mm leaf?

____

____

VARIATION WITHIN A DATA SET

Measures of central tendency don't fully describe variation within a data set. Examine the two small sets listed in table 1.2.

Notice that the mean and the median are informative, but they do not describe variation in the data. The stream fish data set has considerably more variation even though the mean is the same as for the pond fish data set. Variation is best quantified by range, variance, standard deviation, and standard error.

TABLE 1.2

TWO SAMPLE DATA SETS WITH DIFFERENT LEVELS OF VARIATION

Number of Fish Collected in Five Replicate Seine-Net Samples:

Pond Fish Data Set: 25 28 30 32 35

mean ($\bar{x}$) = 30; median = 30

range __________ variance __________

standard deviation __________ standard error __________

Stream Fish Data Set: 10 20 20 25 75

mean ($\bar{x}$) = 30; median = 20

range __________ variance __________

standard deviation __________ standard error __________

The **range** is the difference between the smallest and the largest values of the data set—the wider the range, the greater the variation. The range of the pond fish data set is 25–35 = 10; the range of the stream fish data set is 10–75 = 65. The mean number of fish per sample is the same for both data sets, but the ranges indicate much more variation among the stream samples. Notice that the range can be artificially inflated by one or two extreme values, especially if only a few samples were taken.

Questions 4

Examine the values for the stream and pond fish data sets. In which data set is the variation most influenced by a single value? ______________________________

____

What is the best way to collect data and prevent a single sample from skewing the measures of central tendency and variation? ______________________________

____

Could two samples have the same mean but different ranges? Explain. ______________________________

____

Could two samples have the same range but different means? Explain. ______________________________

____

Variance measures *how* data values vary about the mean. Variance is much more informative than the range, and is easy to calculate (see the following example). First, calculate the mean. Second, calculate the deviation of each sample from the mean. Third, square each deviation. Then sum the deviations. The summation is called the *sum of squared deviations* (or *sum of squares*). Finally, divide the sum of squared deviations by the number of data points minus one to calculate the variance (S^2). The example uses the pond fish data set (table 1.2). Record the calculated values in table 1.2.

Number of Pond Fish Collected x_i	Mean $\bar{x}$	Deviation from the mean $(x_i - \bar{x})$	(Deviation)^2 $(x_i - \bar{x})^2$
25	30	−5	25
28	30	−2	4
30	30	0	0
32	30	2	4
35	30	5	25

Sum of squared deviations = 58
Variance = 14.5

The formulae for the sum of squared deviations and the variance are:

$$\text{sum of squared deviations} = \sum_{i=1}^{N}(x_i - \bar{x})^2 = 58.0$$

$$\text{variance} = \frac{\textit{sum of squared deviations}}{N-1} = 58/4 = 14.5$$

where

N = total number of samples
$\bar{x}$ = the sample mean
x_i = measurement of an individual sample

This formula for sum of squared deviations is really quite simple. The formula $(x_i - \bar{x})^2$ is the squared deviation from the mean for each value. The summation sign ($\sum_{i=1}^{N}$) means to sum all the squared deviations from the first one ($i = 1$) to the last one ($i = N$). The sum of squared deviations (58) divided by the number of samples minus one $(5 - 1 = 4)$ equals the variance. The variance for these data is 58/4 = 14.5.

Variance is a good measure of the dispersion of values about the mean. A second, and more commonly cited, measure of variation is the standard deviation. The **standard deviation** (S) equals the square root of the variance. For our example data set:

$$\text{standard deviation (S)} = \sqrt{\textit{variance}} = \sqrt{14.5} = 3.8$$

Standard deviation is usually reported with the mean in statements such as, "The mean number of fish per sample was 30 ± 3.8." The standard deviation helps us understand the spread of values in a data set. For normal distributions of measurements, the mean ± 1 SD includes 68% of the measurements. The mean ± 2 SD includes 95% of the measurements (fig. 1.1).

Another useful measure of the spread of data about a mean (i.e., variation) is the **standard error** ($S_{\bar{x}}$). This value measures the error from having a limited sample size (N). Clearly, a small sample (N) has more sampling error than does a large sample. The term "sampling error" doesn't mean that we have done something wrong. But we must document sampling error to use later in our calculations of confidence in the sample mean.

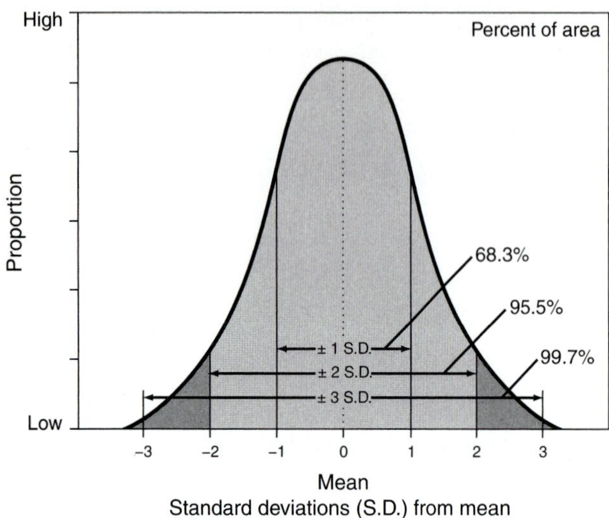

Figure 1.1
Normal distribution graph.

Standard error is calculated as:

$$S_{\bar{x}} = S/\sqrt{N} = 1.70$$

where

S = standard deviation
N = total number of samples

Procedure 1.3

Calculate four measures of variation.

1. Complete table 1.2 by calculating the range, variance, standard deviation, and standard error of the stream fish data set.

Question 5

The range indicates greater variation in the stream fish data set. Do the other three measures of variation also indicate greater variation? ________________

FREQUENCY DISTRIBUTION

Some data sets are better understood if they are displayed in a graph called a **frequency distribution** (fig. 1.2). Frequency distributions summarize data at a glance and reveal subtleties that might not be apparent in calculations of central tendency. The abscissa (*x* axis) is plotted as *Data Class* and the ordinate (*y* axis) as *Frequency of Occurrence* in each Data Class. The raw data for the frequency distribution in figure 1.2 are below the graph. The shape of the curve reveals the nature of variation in the data set. A broad and flat curve reveals high variation. A narrow and high-peak curve reveals less variation.

Frequency distributions often have gradually tapering "tails" of frequencies toward each end of the curve (fig. 1.2). These tails produce a "bell-shaped" curve called a **normal distribution**. A normal distribution is common for ecological data—most values are near the mean and fewer values are at the extremes. Many variables are normally distributed, and many statistical tests used by ecologists assume that the variable is normally distributed.

Procedure 1.4

Examine a frequency distribution of heights.

1. Examine the data presented in figure 1.2 for the height of 119 female college students.

Questions 6

Are the heights distributed as you expected? How so?

____

____

Do you see evidence of central tendency in this data set?

____

Do the data appear normally distributed? ________________

____

In what way do the data deviate from normality? ________________

____

____

2. Examine the mean, median, and mode provided for the data set in figure 1.2.

Question 7

One criterion for a normal distribution is that the mean, median, and mode are equal. Are they equal for the data in figure 1.2? ________________

____

3. Calculate and record the variance, standard deviation, and standard error for the data. The mean and sum of squared deviations are provided to speed your calculations.

Procedure 1.5

Calculate and graph the frequency distribution for a data set for mosquito larvae occurrence in tree-hole cavities.

1. Some mosquito species lay their eggs in tree-hole cavities that hold small volumes of water (< 1 L). Examine the data in figure 1.3 showing the densities of mosquito larvae (number per liter) found in 108 tree-hole cavities.
2. Plot a frequency distribution in the graph provided in figure 1.3.
3. Calculate and record in figure 1.3 the mean, median, mode, variance, standard deviation, and standard error for this data set.

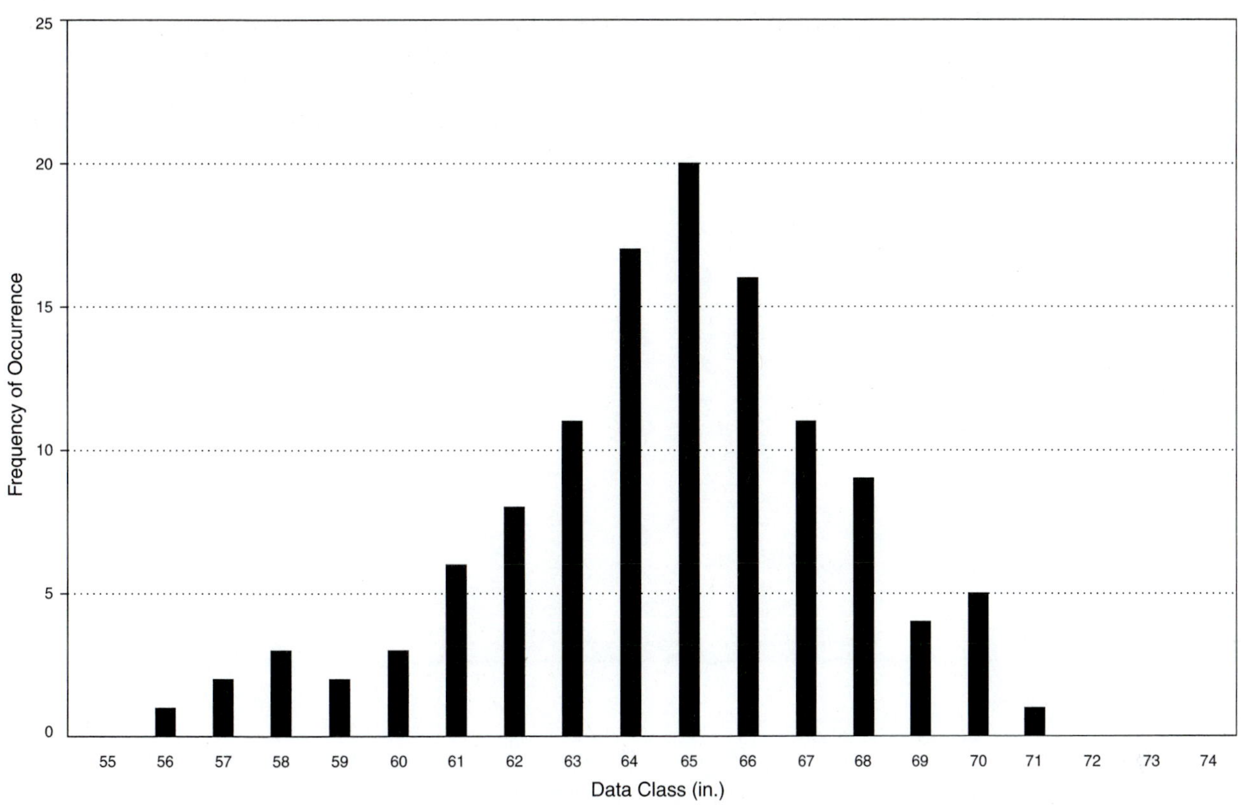

Raw Data (height in inches)							Data Class	Frequency of Occurrence
							Height (in.)	Number of Students
56	62	64	65	66	68		55	0
57	62	64	65	66	68		56	1
57	62	64	65	66	68		57	2
58	62	64	65	66	68		58	3
58	62	64	65	66	68		59	2
58	63	64	65	66	68		60	3
59	63	64	65	66	68		61	6
59	63	64	65	66	68		62	8
60	63	64	65	66	68		63	11
60	63	64	65	67	69		64	17
60	63	64	65	67	69		65	20
61	63	64	65	67	69		66	16
61	63	64	65	67	69		67	11
61	63	65	66	67	70		68	9
61	63	65	66	67	70		69	4
61	63	65	66	67	70		70	5
61	64	65	66	67	70		71	1
62	64	65	66	67	70		72	0
62	64	65	66	67	71		73	0
62	64	65	66	67			74	0
N = 119	Mean = 64.6	Median = 65	Mode = 65	Sum of Squared Deviations = 1078.6		Variance =	Std. Dev. =	Std. Error =

Figure 1.2

A frequency distribution of 119 height measurements of college-aged women.

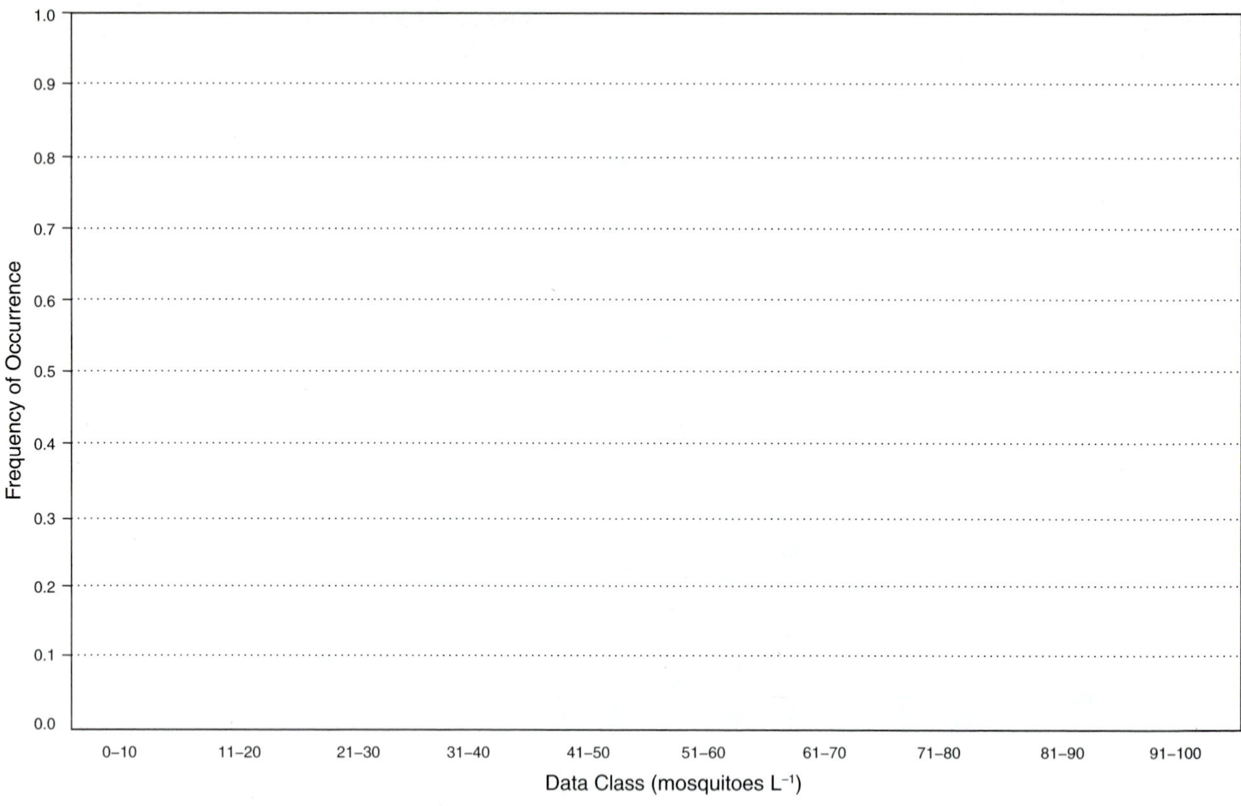

Raw Data (mosquitoes L^{-1})									Data Class	Frequency of Occurrence
									Mosquitoes L^{-1}	Number of Cavities
68	0	50	12	42	24	22	11		0–10	36
24	15	8	29	5	2	99	38		11–20	25
58	56	21	11	32	18	7	23		21–30	17
54	4	26	93	13	29	10	30		31–40	9
3	15	10	1	14	26	0	19		41–50	7
10	5	51	3	2	9	5	17		51–60	6
13	19	71	60	20	20	1	0		61–70	2
7	49	73	17	63	48	14	1		71–80	4
11	44	28	75	30	2	8	12		81–90	0
36	13	17	9	27	6	46	6		91–100	2
8	4	16	53	16	15	41				
5	22	35	9	7	34	37				
19	21	18	74	33	3	8				
39	4	25	5	2	28	31		Mean =	Median =	Mode =

Figure 1.3

Survey of the density (mosquito larvae L^{-1}) of the mosquito, *Aedes triseriatius*, occurring in tree-hole cavities ($N = 108$).

Questions 8
Is this variable normally distributed? ________________

____

How many values were greater than the mean? How many were less? ________________

____

What value best describes the central tendency of this data set? ________________

____

Procedure 1.6

Collect an original data set and calculate its measures of central tendency and variation.

1. Follow your instructor's directions to gather an original data set.
2. Record the raw data in figure 1.4.
3. Calculate and record in figure 1.4 the mean, median, mode, variance, standard deviation, and standard error for these data.

Questions 9
Is the variable in your original data set normally distributed?

____

____

How many values were greater than the mean? ________

How many were less? ________________

What value best describes the central tendency of this data set? ________________

____

POPULATION MEANS AND CONFIDENCE INTERVALS

Population statistics can never be known exactly because we only measure samples of the population, not every member. Therefore, values such as the population mean (μ) must be estimated by the sample mean ($\bar{x}$). If variation is low, then we have high confidence in the sample mean as an estimator of the population mean.

A **confidence interval** is a range of values within which the true population mean occurs with a particular probability. Ecologists usually express their sample means with **95% level of confidence**, also called a **95% confidence interval**. For example, a sample mean ($\bar{x}$) may be 64.6 cm (see figure 1.2 for a sample of height measurements for college-aged women). After calculations, we are 95% confident that the *population* mean lies between 64.0 and 65.1.

The 95% confidence interval surrounding a *population* mean is calculated as:

$$\mu = \bar{x} \pm t_{0.05}(S_{\bar{x}})$$

where

μ = population mean
$\bar{x}$ = sample mean
$t_{0.05}$ = value from student's t table at the 95% confidence level
$S_{\bar{x}}$ = standard error

The value of $t_{0.05}$ is selected from a student's t table available in most statistics textbooks. The appropriate student's t value is determined by the degrees of freedom ($N - 1$) and the confidence level, which in this case is 95% (= 0.05 probability that the population mean occurs outside the range). For example, consult a student t table and you will find that the appropriate $t_{0.05}$ value for a sample of 30 ($N = 30$) and for a 95% confidence interval is 2.045.

To calculate the 95% confidence interval using the oak leaf measurements in table 1.1, first calculate the mean and standard error (see table 1.1).

$\bar{x}$ = mean = _____
$S_{\bar{x}}$ = standard error = _____
$N = 20$
DF = degrees of freedom = $(N - 1) = 19$

A student's t table shows that the critical value of $t_{0.05}$ is 2.09.

Therefore, the 95% confidence intervals surrounding the mean of 20 oak leaf samples is:

$$\mu = \bar{x} \pm 2.09(S_{\bar{x}}) = \_\_\_\_\_\_$$

With this confidence interval, we can state that there is a 95% probability that the population mean of the oak leaves is somewhere between _____ and _____.

Procedure 1.7

Calculate the confidence intervals for example and original data sets.

1. Calculate the 95% confidence interval for the population mean of stream fishes per seine haul from table 1.2.

   $\mu = \bar{x} \pm t_{0.05}(S_{\bar{x}}) = \_\_\_\_ \pm \_\_\_\_$

2. Calculate the 95% confidence interval for the population mean of pond fishes per seine haul from table 1.2.

   $\mu = \bar{x} \pm t_{0.05}(S_{\bar{x}}) = \_\_\_\_ \pm \_\_\_\_$

3. Calculate the 95% confidence interval for the population mean height of female students whose sample is presented in figure 1.2.

   $\mu = \bar{x} \pm t_{0.05}(S_{\bar{x}}) = \_\_\_\_ \pm \_\_\_\_$

Raw Data									Data Class	Frequency of Occurrence
								Mean =	Median =	Mode =
								Variance =	Std. Dev. =	Std. Error =

Figure 1.4
Original data set for analysis.

Questions for Further Thought and Study

1. Replicate samples are central to good experimental design. How would the frequency distribution of heights in figure 1.2 differ if only five or six measurements were made?

2. Do you suspect that any biological variables have a perfectly normal distribution? Why or why not?

3. What is the relationship between variation in a data set and the width of the confidence intervals surrounding the estimate of the population mean?

The Process of Science

2

Objectives

As you complete this lab exercise you will:

1. Define *science* and understand the logic and sequence of the scientific method.
2. Develop productive observations, questions, and hypotheses about the natural world.
3. Calculate the range, mean, and standard deviation for a set of replicate measurements.
4. Design and conduct a controlled experiment to test a null hypothesis.

The term *science* brings to mind different things to different students. To some students, science is a book. To others, it's a microscope, a dissected frog, or a course that you take. In fact, it's none of these things. A good definition of **science** for biological research is *the orderly process of posing and answering questions about the natural world through repeated and unbiased experiments and observations*. This definition emphasizes that science is a *process* rather than a book, or a course, or a list of facts. Science is not a "thing." It's a way of doing things—a way of learning and knowing about the natural world (fig. 2.1).

Our definition also emphasizes that people do science by *asking questions* and then *doing experiments* to answer those questions. Asking questions and being curious are part of human nature, and science is a human activity. Like any human task, it takes practice to do science effectively.

Finally, our definition emphasizes that science is a tool for learning about the *natural world*. It is ineffective for moral choices, ethical dilemmas, and untestable ideas. For example, the scientific method cannot tell us if pollution is good or bad. It can tell us the environmental *consequences* of pollution, but whether these consequences are "good" or "bad" is a judgment that we make based on our values, not on science. Although this is an important limitation of the scientific method, science remains one of the most powerful ways of learning about our world.

Figure 2.1

Science is the process of learning about the natural world. Gathering repeated and unbiased measurements (data) is the engine for testing hypotheses and answering questions. This student is recording the presence or absence of pollinators to answer questions about the daily timing of pollination.

Question 1

What practices besides science are used among world cultures to address questions of the natural world? ______________

____

Questioning and testing are a part of science that enable us to systematically sift through natural variation to find underlying patterns. The natural world includes much variation, and learning biology would be relatively easy if simple observations accurately revealed patterns of the natural world. But they usually don't—nature is too complicated to rely solely on simple observation. We certainly can learn much about our environment by looking around us. But, casual observations are often biased and misleading because nature varies from time to time and from organism to organism. Biologists need a structured and repeatable process for testing ideas about the variation in nature. Science is that process.

Question 2
What factors might cause variation in measurements such as the heights of 10-year-old pine trees, or the kidney filtration rates of 10 lab mice? ______________________

____

The process of science deals with variation through an organized sequence of steps that maintain as much objectivity and repeatability as possible. Although these loosely organized steps, sometimes called the **scientific method**, vary from situation to situation, they are remarkably effective for research and problem solving. Typical steps in the process of science are:

- Make observations
- Pose and clarify testable questions
- Formulate hypotheses
- Conduct experiments to gather data
- Quantify and summarize the data
- Test the hypotheses
- Answer the questions and make conclusions

Figure 2.2
Pill bugs are excellent experimental organisms for testing hypotheses about microenvironments under logs and within leaf litter. They are crustaceans, not insects. Unlike most crustaceans, they are terrestrial rather than aquatic.

DEVELOPMENT OF OBSERVATIONS, QUESTIONS, AND HYPOTHESES

Make Insightful Observations

Good scientists make insightful observations. But that's not as easy as it seems. Consider these two observations:

> Observation 1: Fewer elk live in Yellowstone National Park than in the past.
>
> Observation 2: The density of elk in Yellowstone National Park has declined during the consecutive dry years since the reintroduction of the native wolf population.

Which of these two observations is the strongest and most useful? Both of them may be true, but the second one is much more insightful because it provides a context to the observation that elk populations are declining. It suggests a relevant factor—the introduction of the wolf population—as a productive topic for investigation. It also suggests a relationship between the density of elk and variation in the local environment.

For the remainder of this exercise you will simultaneously develop questions, hypotheses, and designs for two experiments—an experiment involving yeast nutrition (see Worksheet 1 on page 18) and an experiment investigating food preferences for pill bugs (see Worksheet 2, page 19).

Procedure 2.1

Make productive observations.

1. Consider the following two observations.

   Observation 1: Fungi often grow on leftover food.

   Observation 2: Fungi grow on leftover bread more than on leftover meat.

2. Which of these observations is most useful for research? Why? ______________________

   ____

3. Insert the more insightful observation in Worksheet 1.
4. Pill bugs (sometimes called *roly-poly bugs*) are good model organisms for research. They often find food and shelter where fungi are decomposing leaf litter (fig. 2.2). We may be interested in whether pill bugs are attracted to leaves or to fungi growing on the leaves' surface. Consider this observation.

   Observation 1: Pill bugs often hide under things.

5. Propose a more productive observation for a study of pill bug feeding.

   Observation 2: ______________________

   ____

6. Record Observation 2 in Worksheet 2 on page 19. You may revise this later.

Pose and Clarify Testable Questions

Productive observations inspire questions. Humans think in terms of questions rather than abstract hypotheses or numbers. Phrasing a good question takes practice and experience, and the first questions that capture our attention are usually general. For example, "**Which nutrients can yeast most**

readily metabolize?" is a general question that expands the observation posed in Procedure 2.1. This question is broad and the type of question that we ultimately want to understand. Record this as General Question in Worksheet 1.

Broad questions are important, but they are often vague. Therefore, scientists usually refine and subdivide broad questions into more specific ones. For example, a more specific question is "**What classes of biological molecules are most readily absorbed and metabolized by yeast?**" Record this as Specific Question 1 in Worksheet 1.

A further clarification might be "**Does yeast absorb and metabolize carbohydrates better than it absorbs and metabolizes proteins?**" This is a good, specific question because it clearly refers to organisms, processes, and likely variables. It also suggests a path for investigation—that is, it suggests an experiment. Record this as Specific Question 2 in Worksheet 1.

Procedure 2.2

Pose and refine questions.

1. Examine these two questions.

   Question 1: Do songbird populations respond to the weather?

   Question 2: Do songbirds sing more often in warm weather?

2. Which of these questions is the most useful for further investigation? Why? ________________

   ____

3. Examine the following general question, and record it in Worksheet 2.

   General Question: What influences the distribution of pill bugs?

4. Propose a specific question that refers to the food of pill bugs as a variable, and record it here and in Worksheet 2. You may revise this later.

   Specific Question 1 ________________________

   ____

5. Propose a more specific question about pill bugs eating leaves, as opposed to pill bugs eating fungi growing on leaves. Record it here and in Worksheet 2. You may revise this later.

   Specific Question 2 ________________________

   ____

Formulate Hypotheses

Well-organized experiments require that questions be restated as testable hypotheses. A **hypothesis** is a statement that clearly states the relationship between biological variables. A good hypothesis identifies the organism or process being investigated, identifies the response variables and treatment variables, and implies how they will be compared. A hypothesis is a statement rather than a question, and your data analysis ultimately determines whether you reject your hypothesis or accept it (or more formally stated, fail to reject your hypothesis). Accepting a hypothesis does not necessarily mean that it is true. More specifically, it means that you do not have enough evidence to reject it.

An example hypothesis is:

> The mean number of eggs produced per clutch by eagles nesting within 10 km of the coast of Alaska is not significantly different from the mean number produced by eagles nesting more than 10 km from the coast.

Hypotheses are either accepted or rejected. There are no "partial truths" or "middle ground." This may seem like a rather coarse way to deal with questions about subtle natural processes, but using controlled experiments to either accept or reject a hypothesis is proven and effective. The heart of science is gathering and analyzing experimental data that lead to rejecting or accepting hypotheses relevant to the questions we want to answer.

In this exercise you are going to do science as you investigate yeast nutrition and then experiment with food choice by pill bugs. As yeasts ferment their food, CO_2 is produced as a byproduct. Therefore, you can measure the growth of yeast by the production of CO_2 (fig. 2.3).

A hypothesis related to our question about yeast growth might be:

> H_0: CO_2 production by yeasts that were fed sugar is not significantly different from the CO_2 production by yeasts that were fed protein.

A related alternative hypothesis can be similarly stated:

> H_a: Yeasts produce more CO_2 when fed sugar than when fed protein.

The first hypothesis (H_0) is a **null hypothesis** because it states that there is *no difference*. This is the most common way to state a clear and testable hypothesis. Researchers find it more useful to associate statistical probabilities with null hypotheses rather than with alternative hypotheses. It is usually more appropriate to accept or reject a hypothesis with statistics when the hypothesis proposes no effect rather than when there is an effect. Your instructor may elaborate on why researchers test null hypotheses more effectively than alternative hypotheses. Record the null hypothesis in Worksheet 1.

A null hypothesis should be testable and well written. In our experiment, the null hypothesis (1) specifies yeast as the organism, population, or group that we want to learn about; (2) identifies CO_2 production as the variable being measured; and (3) leads directly to an experiment to evaluate treatment variables and compare means of replicated measurements.

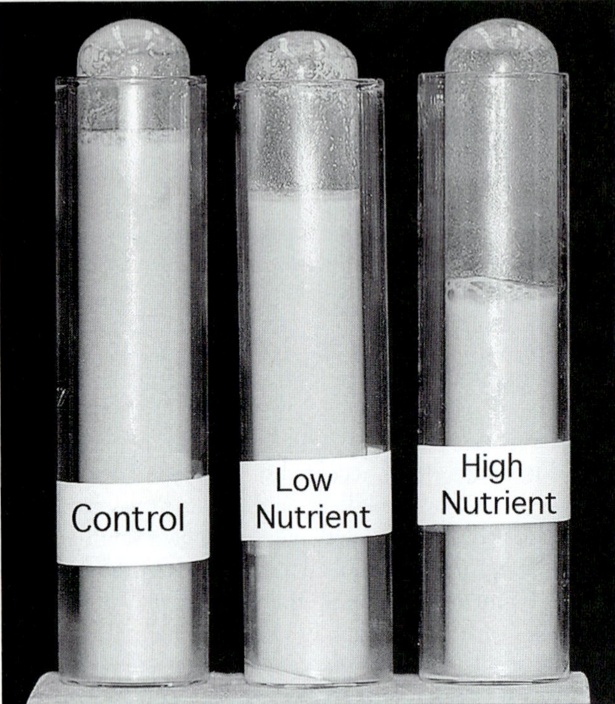

Figure 2.3

These tubes of yeast are fermenting nutrients in solution. The CO_2 produced by the yeasts accumulates in the inverted test tube and indicates the yeasts' rate of metabolism. From right to left, the tubes have abundant nutrients, low nutrients, and a control with no added nutrients, respectively.

Procedure 2.3

Formulate hypotheses.

1. Examine the following two hypotheses.

   Hypothesis 1: Songbirds sing more when the weather is warm.

   Hypothesis 2: The number of bird songs heard per hour during daylight temperatures above 80°F is not significantly different from the number heard per hour at temperatures below 80°F.

2. Which of these hypotheses is the most useful for further investigation? Why? ____________________

   ____

3. Which of these hypotheses is a null hypothesis? Why? ____________________

   ____

4. Examine the following hypothesis.

   Hypothesis 1: Pill bugs prefer leaves coated with a thin layer of yeast.

5. Propose a more effective null hypothesis. Be sure that it is null, that it is testable, and that it includes the variable you will control in an experiment.

   Hypothesis 2 (H_0): ____________________

   ____

6. Record your null hypothesis in Worksheet 2.

EXPERIMENTATION AND DATA ANALYSIS: YEAST NUTRITION

Gather Experimental Data

To test our hypothesis about yeast growth we must design a controlled and repeatable experiment. The experiment suggested by our specific question and hypothesis involves offering a sugar such as glucose to one population of yeast, offering protein to another population of yeast, and then measuring their respective growth rates. Fortunately, yeast grows easily in test tubes. As yeast grows in a closed, anaerobic container it produces CO_2 in proportion to how readily it uses the available food. You can easily measure CO_2 production by determining the volume of CO_2 that accumulates at the top of an inverted test tube.

Experiments provide data that determine if a null hypothesis should be accepted or rejected. A well-designed experiment links a biological response to different levels of the variable being investigated. In this case, the biological response is CO_2 production indicating growth. The levels of the variable are sugar and protein. These levels are called **treatments**, and in our experiment they include glucose, protein, and a control. The **treatment variable** being tested is the type of food molecule (i.e., protein, sugar), and the **response variable** is the CO_2 production that indicates yeast growth.

A good experimental design compensates for natural variation. It should (1) include replications; (2) test only one treatment variable; and (3) include controls. **Replications** are repeated measures of each treatment under the same conditions. Replications effectively deal with naturally occurring variation. Usually, the more replications, the better. Your first experiment today includes replicate test tubes of yeast, each treated the same. Good design also tests only one treatment variable at a time.

Good experimental design also requires **controls**. Controls are standards for comparison, and they verify that the biological response we measure is a function of the variable being investigated and nothing else. They are replicates with all of the conditions of an experimental treatment *except the treatment variable*. For example, if the treatment is glucose dissolved in water, then a control has only water (i.e., it lacks glucose, the treatment variable). This verifies that the response is to glucose and not to the solvent.

Procedure 2.4

Conduct an experiment to determine the effects of food type on yeast growth.

1. Label 12 test tubes as C1–C4, G1–G4, and P1–P4. See Worksheet 1, page 18.
2. Add 5 mL of water to test tubes C1–C4. These are control replicates.
3. Add 5 mL of 5% glucose solution to test tubes G1–G4. These are replicates of the glucose treatment.
4. Add 5 mL of 5% protein solution to test tubes P1–P4. These are replicates of the protein treatment.
5. Completely fill the remaining volume in each tube with the yeast suspension provided.
6. For each tube, slide an inverted, flat-bottomed test tube down over the yeast-filled tube. Hold the yeast-filled tube firmly against the inside bottom of the cover tube and invert the assembly. Your instructor will demonstrate how to slip this slightly larger empty tube over the top of each yeast tube and invert the assembly. If done properly, no bubble of air will be trapped at the top of the tube of yeast after inversion.
7. Place the tubes in a rack and incubate them at 50°C for 40 minutes.
8. After 40 minutes, measure the height (mm) of the bubble of accumulated CO_2. Record your results in Worksheet 1.

Analyze the Experimental Data

Analysis begins with summarizing the raw data for biological responses to each treatment. The first calculation is the **mean** ($\bar{x}$) of the response variable (mm CO_2) for replicates of each treatment and controls. The mean represents the central tendency of all measurements (replicates) of the response variable. Later, the mean of each treatment will be compared to determine if the treatments had different effects.

The second step in data analysis is to calculate variation within each set of replicates. The simplest measure of variation is the **range**, which is the highest and lowest values in a set of replicates. A wide range indicates much variation. The **standard deviation (SD)**, another informative measure of variation, summarizes the variation just as the range does, and the standard deviation is less affected by extreme values. Refer to the description in Exercise 1 for calculating the standard deviation.

Procedure 2.5

Quantify and summarize the data.

1. Examine your raw data in Worksheet 1.
2. Calculate and record in Worksheet 1 the mean of the response variable (CO_2 production) for the four control replicates. To calculate the means for the four replicates, sum the four values and divide by 4.
3. The CO_2 production for each glucose and protein replicate must be adjusted with the control mean. This ensures that the final data reflect the effects of only the treatment variable and not the solvent. Subtract the control mean from the CO_2 production of each glucose replicate and each protein replicate, and record the results in Worksheet 1.
4. Record in Worksheet 1 the range of adjusted CO_2 production for the four replicates of the glucose treatment and of the protein treatment.
5. Calculate the mean CO_2 production for the four adjusted glucose treatment replicates. Record the mean in Worksheet 1.
6. Calculate the mean CO_2 production for the four adjusted protein treatment replicates. Record the mean in Worksheet 1.
7. Refer to the description of standard deviation in Exercise 1, and calculate the standard deviation for the four adjusted glucose treatment values, and for the four adjusted protein treatment values. Record the two standard deviations in Worksheet 1.

Test the Hypotheses

Our hypothesis about yeast growth is tested by comparing the mean CO_2 production by yeast that was fed glucose to the mean CO_2 production by yeast that was fed protein. However, simply determining if one mean is higher than the other is not an adequate test because natural variation always makes the two means at least slightly different even if the two treatments have the same effect on yeast growth. Therefore, the means must be compared to determine if the means are not just different, but are **significantly different**. To be significantly different means that the differences between means are due to the treatment, and not just due to natural variation. If the difference is significant, then the null hypothesis is rejected. If the difference is not significant, then the null hypothesis is accepted. Testing for significant differences is usually done with statistical methods.

Statistical methods calculate the probability that the means are significantly different. We will use a simple method to test for a significant difference between the means of our two treatments. We will declare that two means are significantly different *if the 95% confidence intervals surrounding the two means do not overlap*. Review Exercise 1 for the calculations of a 95% confidence interval. This simple criterion will suffice as a test for significant differences. Your instructor may choose to present a more rigorous statistical test for significance.

For example, consider these two means and their 95% confidence intervals:

$\text{mean}_a = \bar{x}_a = 10$ $\mu_{0.05} = 10 \pm 2.5$

$\text{mean}_b = \bar{x}_b = 20$ $\mu_{0.05} = 20 \pm 5$

Are mean_a and mean_b significantly different according to our test for significance? Yes, they are because the confidence interval $7.5 \leftrightarrow 12.5$ does not overlap $15 \leftrightarrow 25$.

Procedure 2.6

Testing the hypothesis.

1. Consider your null hypothesis and the data presented in Worksheet 1.
2. Calculate and record in Worksheet 1 the 95% confidence interval for the glucose mean.
3. Calculate and record in Worksheet 1 the 95% confidence interval for the protein mean.
4. Do the confidence intervals surrounding the two treatment means overlap? Record your answer in Worksheet 1.
5. Are the means for the two treatments significantly different? Record your answer in Worksheet 1.
6. Is your null hypothesis accepted or rejected? Record your answer in Worksheet 1.

Answer the Questions

The results of testing the hypotheses are informative, but it still takes a biologist with good logic to translate these results into answers for our specific and general questions. If your specific questions were well-stated, then answering them based on your experiment and hypothesis testing should be straightforward.

Procedure 2.7

Answer the questions: yeast nutrition.

1. Examine the results of hypothesis testing presented in Worksheet 1.
2. *Specific Question 2* is "Does yeast absorb and metabolize carbohydrates better than it absorbs and metabolizes proteins?" Record your answer in Worksheet 1.
3. Does your experiment adequately answer this question? Why or why not? ____________

   ____
4. *Specific Question 1* is "What classes of biological molecules are most readily absorbed and metabolized by yeast?" Record your best response in Worksheet 1.
5. Does your experiment adequately answer *Specific Question 1*? Why or why not? ____________

   ____

   ____
6. The *General Question* is "Which nutrients can yeast most readily metabolize?" After testing the hypotheses, are you now prepared to answer this *General Question*? Why or why not? ____________

   ____
7. Record your best response to the *General Question* in Worksheet 1.

EXPERIMENTATION AND DATA ANALYSIS: PILL BUG FOOD PREFERENCE

In the previous procedures, you developed and recorded observations, questions, and hypotheses related to pill bug food preference. Pill bugs may be attracted to dead leaves as food, or they may be attracted to fungi growing on the leaves as food. Leaves dipped in a yeast suspension can simulate fungi growing on leaves. Use the following procedures as a guide to the science of experimentation and data analysis to test your hypothesis recorded in Worksheet 2.

Procedure 2.8

Design an experiment to test pill bug food preference.

1. Design an experiment to test your hypothesis in Worksheet 2 about pill bug food preference. To do this, specify:

   Experimental setup ____________

   ____

   ____

   Treatment 1 to be tested ____________

   Treatment 2 to be tested ____________

   Control treatment ____________

   Response variable ____________

   Treatment variable ____________

   Number of replicates ____________

   Means to be compared ____________
2. Conduct your experiment and record the data in Worksheet 2.
3. Analyze your data. Record the control means and the adjusted treatment means in Worksheet 2.
4. Calculate the ranges and standard deviations for your treatments. Record them in Worksheet 2.
5. Test your hypothesis. Determine if the null hypothesis should be accepted or rejected. Record the results in Worksheet 2.

Procedure 2.9

Answer the questions: pill bug food preference.

1. Examine the results of your hypothesis testing in Worksheet 2.
2. Record your answer to *Specific Question 2* in Worksheet 2. Does your experiment adequately answer this question? Why or why not? ______________

   ____
3. Record your best response to *Specific Question 1* in Worksheet 2. Does your experiment adequately answer this question? Why or why not? ______________

   ____
4. After testing the hypotheses, are you now prepared to answer your *General Question* "What influences the distribution of pill bugs?" Why or why not?

   ____

   ____

   ____
5. Record your response to the *General Question* in Worksheet 2.

Worksheet 1: Process of Science: Nutrient Use in Yeast

OBSERVATION ______________________________

QUESTIONS

General Question: ______________________________

Specific Question 1: ______________________________

Specific Question 2: ______________________________

HYPOTHESIS H_0: ______________________________

EXPERIMENTAL DATA: Nutrient Use in Yeast

TREATMENTS						TREATMENTS MINUS CONTROL $\bar{x}$	
Rep.	Control CO_2 Production (mm)	Rep.	Glucose CO_2 Production (mm)	Rep.	Protein CO_2 Production (mm)	Glucose CO_2 Production Adjusted for the Control $\bar{x}$	Protein CO_2 Production Adjusted for the Control $\bar{x}$
C1	____	G1	____	P1	____	____	____
C2	____	G2	____	P2	____	____	____
C3	____	G3	____	P3	____	____	____
C4	____	G4	____	P4	____	____	____
Control $\bar{x}$ = ____						Glucose $\bar{x}$ = ____	Protein $\bar{x}$ = ____
						Glucose range = ____ – ____	Protein range = ____ – ____
						Glucose SD = ____	Protein SD = ____

TEST HYPOTHESIS

Glucose treatment: $\mu_{glucose}$ ± 95% confidence interval = $\mu_{glucose}$ ± ______

Protein treatment: $\mu_{protein}$ ± 95% confidence interval = $\mu_{protein}$ ± ______

Do the 95% confidence intervals surrounding the means of the two treatments overlap? Yes ______ No ______

Are the means for the two treatments significantly different? Yes ______ No ______

Is the null hypothesis accepted? ______ or rejected? ______

ANSWER QUESTIONS

Answer to Specific Question 2: ______________________________

Answer to Specific Question 1: ______________________________

Answer to General Question: ______________________________

Worksheet 2: Process of Science: Food Preference by Pill Bugs

OBSERVATION ______

QUESTIONS

General Question: ______

Specific Question 1: ______

Specific Question 2: ______

HYPOTHESIS H_0: ______

EXPERIMENTAL DATA: Food Preference by Pill Bugs

TREATMENTS						TREATMENTS MINUS CONTROL $\bar{x}$	
Rep.	Control	Rep.	Treatment 1	Rep.	Treatment 2	Treatment 1 Adjusted for the Control $\bar{x}$	Treatment 2 Adjusted for the Control $\bar{x}$
1	____	1	____	1	____	____	____
2	____	2	____	2	____	____	____
3	____	3	____	3	____	____	____
4	____	4	____	4	____	____	____
Control $\bar{x}$ = ____						Treat 1 $\bar{x}$ = ____	Treat 2 $\bar{x}$ = ____
						Treat 1 Range = ___ – ___	Treat 2 Range = ___ – ___
						Treat 1 SD = ____	Treat 2 SD = ____

TEST HYPOTHESIS

For Treatment 1: $\mu_{\text{treat 1}}$ ± 95% confidence interval = $\mu_{\text{treat 1}}$ ± ______

For Treatment 2: $\mu_{\text{treat 2}}$ ± 95% confidence interval = $\mu_{\text{treat 2}}$ ± ______

Do the 95% confidence intervals surrounding the means of the two treatments overlap? Yes ______ No ______

Are the means for the two treatments significantly different? Yes ______ No ______

Is the null hypothesis accepted? ______ or rejected? ______

ANSWER QUESTIONS

Answer to Specific Question 2: ______

Answer to Specific Question 1: ______

Answer to General Question: ______

Questions for Further Thought and Study

1. Newspaper articles often refer to a discovery as "scientific," and claim it was proven "scientifically." What is meant by this description?

2. Experiments and publications in science are usually reviewed by other scientists. Why is this done?

3. Have all of our discoveries and understandings about the natural world been the result of applying the scientific method? How so?

4. Suppose that you hear that two means are *significantly* different. What does this mean?

5. Can means be different but not significantly different? Explain your answer.

Soil Analysis

3

Objectives

As you complete this lab exercise you will:

1. Examine and compare soil horizons.
2. Measure the vertical temperature gradient, soil pH, density, dry weight, organic content, and moisture content for samples along a soil profile.
3. Measure the content of major nutrients in selected soil horizons.
4. Determine the particle size distribution of a soil sample.

Terrestrial plants and animals depend on soil as more than a convenient place to live. It's a storehouse of valuable resources. It's a place to find water, to find valuable nitrogen and phosphorus, and to find food. It's a place to hide, a place to decompose, and a place that changes faster than most people realize. And, like every other aspect of the environment, it varies from place to place. In terrestrial habitats, soils, along with climate, determine the variety and success of all of the residents (fig. 3.1).

Soil is old. It began forming soon after the Earth was formed more than 4.5 billion years ago. Early on, the Earth was a harsh place—"land" was little more than a mixture of igneous, sedimentary, and metamorphic rocks. Slowly these rocks were transformed into soil by glaciers, wind, and rain, and later by the activities of organisms. This conversion of rock into soil is called **weathering**, and it is constant, but slow—so slow that in eastern North America, abundant, weathering rainfall forms only 2 cm of topsoil every 200 years! Soil is a fragile and valuable resource that needs our protection.

Figure 3.1
Soil's origin from weathered rocks and subsequent mixing with organic material typically produces a diverse environment of particles, clumps, detritus, minerals, and microorganisms.

Question 1
Can water actually break rocks? How so?________________

____

SOIL PROFILES AND HORIZONS

A vertical soil profile typically has several layers, called **horizons** (fig. 3.2). A profile is a snapshot of soil structure, but soil is more dynamic than it first appears. Climate affects the rate of weathering of parent materials, the rate of leaching of organic and inorganic substances, the rate of erosion and transport of mineral particles, and the rate of decomposition of organic matter. These processes produce a soil profile's distinctive horizons.

The **O horizon** is the surface litter covering the soil. It is mainly composed of fallen leaves and is only a few centimeters thick. The deeper parts of the O horizon have highly fragmented and partially decomposed organic matter. Fragmentation and decomposition of the organic matter result

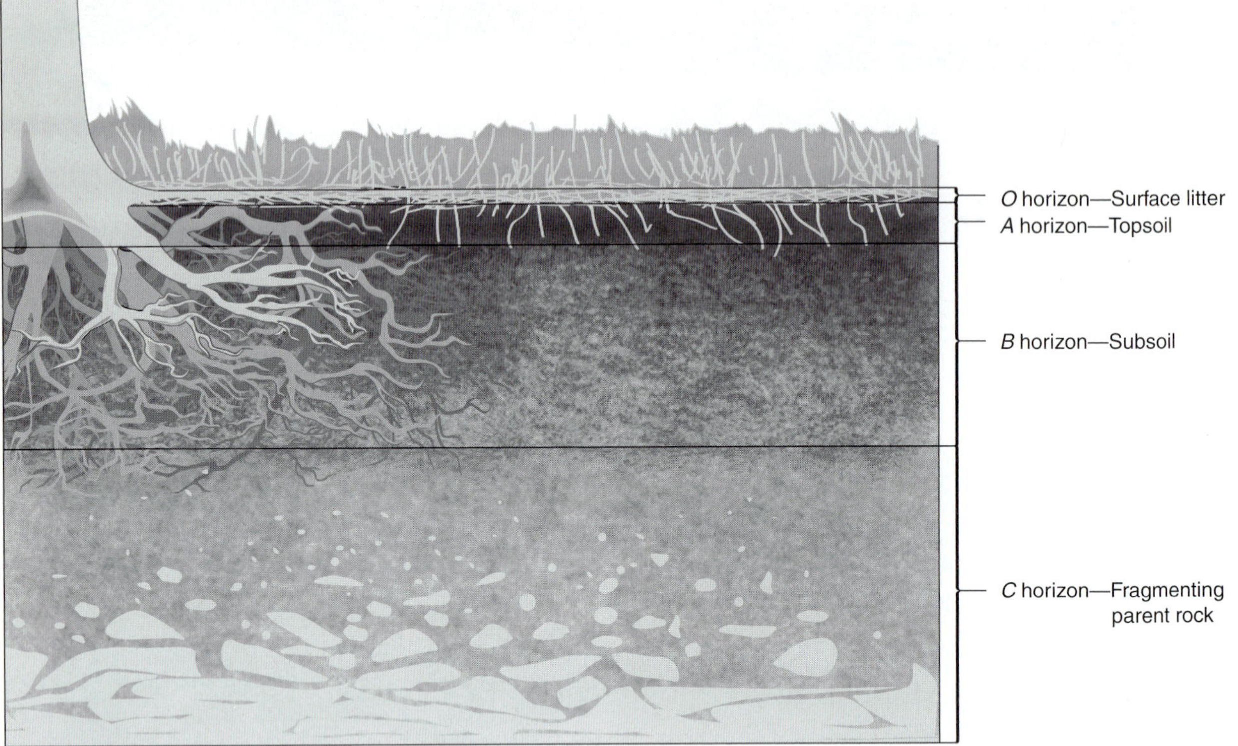

Figure 3.2
Horizons of a verticle soil profile.

from activities of soil organisms, including bacteria, fungi, and animals ranging from nematodes and mites to burrowing mammals. The O horizon merges gradually with the A horizon.

The **A horizon** is **topsoil** and usually extends 10–30 cm below the surface. In most fertile soils, the A horizon has a pH near 7 and contains 10%–15% organic matter, which makes the horizon a dark color. The A horizon contains a mixture of mineral materials such as clay, silt, and sand, and organic material derived from the O horizon.

The **B horizon** has larger soil particles than those in the A horizon and extends 30–60 cm below the soil surface. This horizon contains progressively less organic matter and is therefore lighter in color than the overlying A horizon. In many regions, the B horizon contains large amounts of minerals and clay particles washed by rainfall from the A horizon. Mature roots commonly extend into the B horizon, where minerals accumulate. The B horizon is often called *subsoil*.

The **C horizon** occurs 60–120 cm below the surface and consists of weathering rock subject to the action of frost, water, and the deeper penetrating roots of plants. This horizon usually lacks organic matter and is often called **parent material**, because it is the raw material from which soil forms. The C horizon extends to an underlying and often impenetrable bedrock of igneous, sedimentary, or metamorphic rock.

Regardless of the different properties of their horizons, all soils contain the same five components: mineral particles, decaying organic matter (humus), air, water, and living organisms. Differing amounts of these materials define the soil's properties and the plants it can support.

For this lab exercise, you will measure the primary characteristics of a variety of soil samples. Your instructor will design your class's survey of various sites, horizons, depths, and replications. Each of the following procedures applies to a single sample, but includes tables to record as many as four sets of values for each soil sample. Check with your instructor about how many and which soil samples your group will process. Your instructor has selected two or more communities with contrasting soil types and soil profiles. Notice the differences in the plant communities of each site.

Procedure 3.1

Examine and compare soil horizons.

1. Identify two (or more) local communities with contrasting soil types. Consult with your instructor and with geological survey maps of your area to locate potential sampling sites.

2. Expose a soil profile at each site by taking soil cores, or by digging a hole to reveal a vertical profile about 1 m deep.
3. Measure and record in step 4 the depth (thickness) of each horizon for each site.
4. Site____________________

   Horizon ____________________

   Depth __________ cm

   Thickness __________ cm

   Site ____________________

   Horizon ____________________

   Depth __________ cm

   Thickness __________ cm

   Site ____________________

   Horizon ____________________

   Depth __________ cm

   Thickness __________ cm

   Site ____________________

   Horizon ____________________

   Depth __________ cm

   Thickness __________ cm

Questions 2

The boundaries of horizons are not always distinct. What physical processes blend the boundaries? __________

____

What kinds of organisms and what biotic processes blend soil boundaries?____________________

____

SOIL TEMPERATURE

Soil temperature is more important to the ecology of plants and animals than you might expect. Topsoil is a reasonably good insulator and offers burrowing animals a cool refuge from daytime heat, and a stable, insulated environment during harsh winter freezes. The lower horizons typically have a cooler and narrower temperature range. Insulation from freezing air also keeps deeper soil moisture from freezing. Plants cannot absorb frozen water.

Procedure 3.2

Measure the vertical temperature gradient in a soil profile.

1. Obtain a suitable thermometer—either a dial thermometer with a metal "probe," or a linear thermometer with a metal jacket that can be pushed into the ground. Sophisticated thermistors on long metal probes are sometimes used.
2. The probe of a dial thermometer will provide temperature readings for air, soil surface, and shallow depths. For deeper depths, use a flat, sharp-edged shovel to dig a hole that exposes a vertical profile as deep as possible. Then quickly push the probe horizontally 5–10 cm into the horizon at various depths (preferably every 15 cm).
3. Record in step 4 the temperature profile for a sampling site.
4. Site____________________

   Horizon ____________________

   Depth __________ cm

   Temperature __________ °C

   Site____________________

   Horizon ____________________

   Depth __________ cm

   Temperature __________ °C

   Site____________________

   Horizon ____________________

   Depth __________ cm

   Temperature __________ °C

   Site____________________

   Horizon ____________________

   Depth __________ cm

   Temperature __________ °C

Questions 3

How would you design an experiment to determine the depths of soil that are subject to daily temperature fluctuations? ____________________

____

How might organisms take advantage of the insulating properties of soil? ____________________

____

SOIL pH

Soil fertility is strongly affected by pH. The subtleties of soil chemistry are beyond the scope of this exercise, but acidity can strongly influence nutrient availability. Nutrients must be in solution before plant roots can absorb them easily, and nutrient solubility depends much on pH. Most minerals and nutrients dissolve better in acidic soils than in neutral or basic (alkaline) soils. The pH of soils ranges from roughly 3 (acidic peat bogs) to 10 (dry desert soils). Most farm soils have a pH range of 4.5–9. A pH of 5–7 is optimum for most plants.

Procedure 3.3

Measure soil pH

1. If a soil test kit is available (fig. 3.3), use it to measure pH of a soil sample taken 5–10 cm deep, and another taken 30–40 cm deep. This method relies on the color change of indicator chemicals. Record the values in step 3.
2. If a pH meter is used rather than chemicals in a soil test kit, then collect the two soil samples mentioned in step 1. Mix equal volumes of soil and water in a beaker and measure the pH of the suspension with the pH meter. Record the values in step 3.
3. Site ______________________

   Horizon ______________________

   Depth ______________ cm

   pH ______________

   Site ______________________

   Horizon ______________________

   Depth ______________ cm

   pH ______________

   Site ______________________

   Horizon ______________________

   Depth ______________ cm

   pH ______________

   Site ______________________

   Horizon ______________________

   Depth ______________ cm

   pH ______________

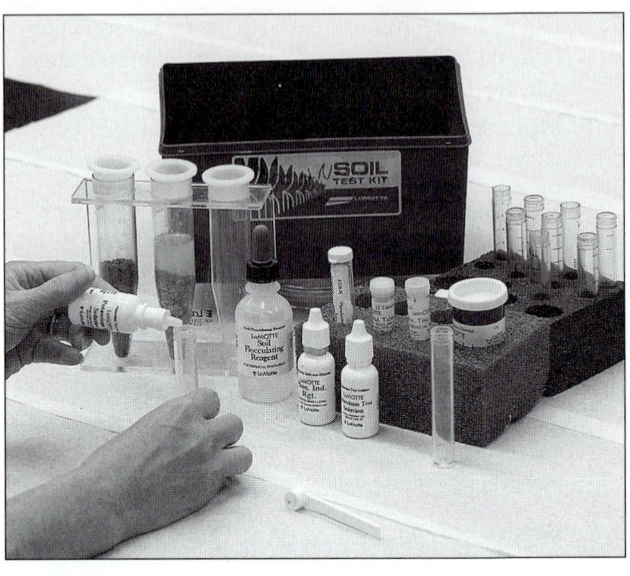

Figure 3.3

This test kit provides the chemicals, mixing tubes, and solutions to detect physical and chemical properties of soil.

Questions 4

Examine some potting soil. What pH would you expect it to have? Why? ______________________

____

What characteristics of potting soil make it a good medium for cultured plants? ______________________

____

SOIL DENSITY

Soil densities can vary considerably. The moisture content, ratio of particle sizes, and organic content all affect density.

Procedure 3.4

Measure soil density.

1. Obtain and weigh a 1-L beaker.
2. Fill the beaker with 1 L of water, and accurately mark the water level on the side of the beaker. Empty and dry the beaker.
3. Fill a 1-L graduated cylinder with 1 L of coarse sand.
4. Use a sharp-edged trowel to dig a soil sample of 400–500 mL and put it in the beaker. Be careful not to compact the soil.
5. Weigh the beaker and soil sample and subtract the weight of the beaker (step 1) to obtain the weight of the soil sample. Record the weight in step 9.

6. Slowly pour sand from the graduated cylinder over and around the soil sample in the beaker until the total volume in the beaker reaches 1 L.
7. Read the remaining volume of sand in the cylinder. The volume of remaining sand equals the volume of the soil sample. Record soil volume in step 9.
8. Calculate and record in step 9 the density of the soil sample as:
   soil density (g L^{-1}) = (weight of soil sample) / (volume of soil sample)
9. Site ______________________

   Horizon ______________________

   Depth ____________ cm

   Soil density ________ g L^{-1}

   Beaker ____________ g

   Soil and beaker ________ g

   Soil ____________ g

   Soil volume ________ mL

   Site ______________________

   Horizon ______________________

   Depth ____________ cm

   Soil density ________ g L^{-1}

   Beaker ____________ g

   Soil and beaker ________ g

   Soil ____________ g

   Soil volume ________ mL

   Site ______________________

   Horizon ______________________

   Depth ____________ cm

   Soil density ________ g L^{-1}

   Beaker ____________ g

   Soil and beaker ________ g

   Soil ____________ g

   Soil volume ________ mL

   Site ______________________

   Horizon ______________________

   Depth ____________ cm

   Soil density ________ g L^{-1}

   Beaker ____________ g

   Soil and beaker ________ g

   Soil ____________ g

   Soil volume ________ mL

Questions 5

Do you notice any places near your study sites where soil has been significantly compacted? ______________________

____

How might compaction of soil affect its oxygen content? Moisture content? Overall plant success? ______________________

____

SOIL MOISTURE CONTENT

Soil retains water. The amount of retained water is proportional to the surface area of the soil's particles—the larger the total surface area, the greater the retention of water. Clay particles are smaller than sand and therefore have a much larger surface area per unit of soil volume than does sand. Indeed, the surface of clay particles in the upper few centimeters of soil in a 2-hectare cornfield equals the surface area of North America. Because clay soils retain much more water than do sandy soils, clays would seem ideal for plant growth. But the small size of clay particles also results in being densely packed—so densely that the clay has low amounts of oxygen, due to small air spaces. This density also retards the penetration of water into the soil (e.g., water penetrates clay about 20 times slower than it penetrates sand). As a result, much of the water that falls on clay soil runs off and is unavailable for plant growth. Tightly packed clay can impede plant growth.

Procedure 3.5

Measure the fresh weight, dry weight, and moisture content of a soil sample.

1. Collect a soil sample (≈ 50 g) and seal it in a pre-weighed bag.
2. Weigh the bag with soil sample and subtract the weight of the bag to calculate the sample's fresh weight. Record this fresh weight in step 8.
3. Transfer the soil to an open, pre-weighted container such as a small aluminum pan or glass dish. Break up any chunks so that the soil dries evenly.
4. Dry the soil for 24 h at 110°C.
5. After drying, use tongs to place the pan and soil in a desiccator until it cools. Do not seal the desiccator jar completely.

6. Weigh the cooled pan with soil. Subtract the weight of the pre-weighed pan and record in step 8 the remaining dry weight of the soil.
7. Calculate and record in step 8 the percent moisture content as:

   % moisture content = 100 · (fresh weight − dry weight) / fresh weight
8. Site ________________________

   Horizon ________________________

   Depth ____________ cm

   Moisture content ______ %

   Collection bag ________ g

   Bag with soil sample _____ g

   Fresh soil __________ g

   Drying pan __________ g

   Pan with dried soil ______ g

   Soil dry weight _______ g

   Site ________________________

   Horizon ________________________

   Depth ____________ cm

   Moisture content ______ %

   Collection bag ________ g

   Bag with soil sample _____ g

   Fresh soil __________ g

   Drying pan __________ g

   Pan with dried soil ______ g

   Soil dry weight _______ g

   Site ________________________

   Horizon ________________________

   Depth ____________ cm

   Moisture content ______ %

   Collection bag ________ g

   Bag with soil sample _____ g

   Fresh soil __________ g

   Drying pan __________ g

   Pan with dried soil ______ g

   Soil dry weight _______ g

   Site ________________________

   Horizon ________________________

   Depth ____________ cm

   Moisture content ______ %

   Collection bag ________ g

   Bag with soil sample _____ g

   Fresh soil __________ g

   Drying pan __________ g

   Pan with dried soil ______ g

   Soil dry weight _______ g

Question 6

Some desert sands receive significant rainfall but only support desert plants. Why? ____________________

____

ORGANIC CONTENT AS ASH-FREE DRY WEIGHT

Humus is the decomposing organic matter in soil. The amount of humus in soil varies; heavily mineralized soils contain 1%–10% humus, whereas organic soils typically contain about 30% humus. Most plants grow best in soil containing 10%–20% humus. The most organic soils are those of swamps and bogs, which may contain more than 90% humus. These soils are usually so acidic that decomposers grow poorly in them. As a result, swamp humus accumulates faster than it is broken down.

The amount of humus in soil affects the soil and plants in several ways:

- Its lightweight and spongy texture increases the water-retention capacity of the soil. Water absorption by humus decreases runoff, thereby slowing erosion.
- Most humus is rich in organic acids and tends to increase nutrient availability.
- Humus swells and shrinks as it absorbs water and later dries. This periodic swelling and shrinking aerates the soil.
- Humus is a reservoir of nutrients for plants. Like time-release vitamins, humus gradually releases nutrients as it is degraded by decomposers.

Procedure 3.6

Measure organic content.

1. Collect a soil sample (≈ 30 g) from a selected horizon and seal it in a plastic bag.

2. Transfer the soil to an open container such as a small aluminum pan or glass dish. Break up any chunks so that the soil dries evenly.
3. Dry the soil for 24 h at 110°C, and homogenize the sample after drying.
4. Weigh and record in step 10 a ceramic crucible to the nearest 0.1 mg. Then add 1–5 g of the oven-dried soil sample.
5. Weigh the filled crucible and subtract the original crucible weight to obtain the oven-dry weight of the sample. Record this weight in step 10.
6. Place the crucible in a muffle furnace and heat to 500°C for 6 h. Do not exceed 500°C. If black, unoxidized material still remains, then heat for two more hours (fig. 3.4).
7. Allow the furnace to cool for 3 h before removing the crucible. Use tongs to remove the crucible and place it in a desiccator to attain room temperature. Do not seal the desiccator completely.
8. Weigh the crucible with the ashed soil to the nearest 0.1 mg and subtract the original weight of the crucible to determine the ash-free dry weight. Record these weights in step 10.
9. Calculate and record in step 10 the percent organic matter as:

   percent organic matter = (oven dry weight − ash free dry weight) / oven dry weight

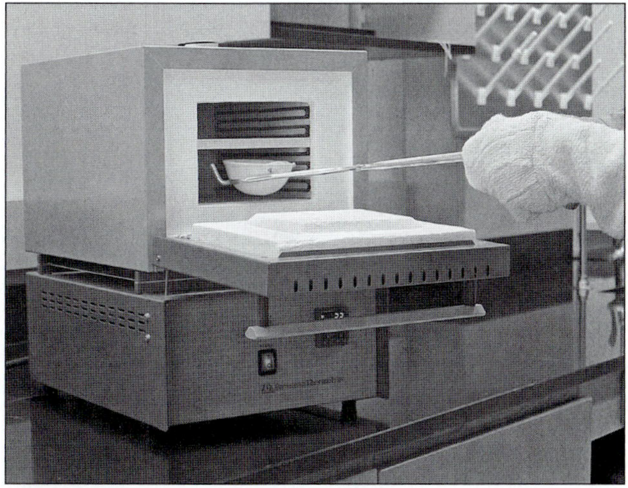

Figure 3.4

Muffle furnaces commonly reach temperatures above 700°F. **Be careful**. High temperatures volatilize organic molecules and leave behind inorganic ash. Samples are typically held in porcelain crucibles that do not melt at high temperatures.

10. Site ____________________

Horizon ____________________

Depth ____________________ cm

Organic matter ____________________ %

Collection bag ____________________ g

Crucible ____________________ g

Crucible with soil ____________________ g

Oven dry soil ____________________ g

Crucible with ashed soil ____________________ g

Ash free dry weight ____________________ g

Site ____________________

Horizon ____________________

Depth ____________________ cm

Organic matter ____________________ %

Collection bag ____________________ g

Crucible ____________________ g

Crucible with soil ____________________ g

Oven dry soil ____________________ g

Crucible with ashed soil ____________________ g

Ash free dry weight ____________________ g

Site ____________________

Horizon ____________________

Depth ____________________ cm

Organic matter ____________________ %

Collection bag ____________________ g

Crucible ____________________ g

Crucible with soil ____________________ g

Oven dry soil ____________________ g

Crucible with ashed soil ____________________ g

Ash free dry weight ____________________ g

Site ____________________

Horizon ____________________

Depth_______________ cm

Organic matter _______________ %

Collection bag _______________ g

Crucible_______________g

Crucible with soil _______________ g

Oven dry soil_______________g

Crucible with ashed soil _______________g

Ash free dry weight_______________g

Questions 7

Did organic content relate to soil color? How so? _______

____

Consider the characteristics of soil discussed so far. How would organic content rank among them as a predictor of plant community success? _______________

____

NUTRIENT CONTENT

Nutrient content obviously determines much about a soil's fertility. Nitrates and phosphates often limit growth, and their concentrations significantly influence a plant community's growth rate as well as which species are persistent and competitive. Nutrient measurements are best done with a soil analysis kit (see fig. 3.3).

Procedure 3.7

Measure the content of major nutrients in selected soil horizons.

1. Obtain a LaMotte soil analysis kit.
2. Select sites to obtain soil samples for comparison, and sample the horizons selected by your instructor.
3. Follow the directions provided in the analysis kit to measure phosphate and nitrate for each soil sample. Record your results in step 4.
4. Site _______________

   Horizon _______________

   Depth_______________ cm

   Phosphate _________mg L^{-1}

   Nitrate _________mg L^{-1}

   Site _______________

   Horizon _______________

   Depth_______________ cm

   Phosphate _________mg L^{-1}

   Nitrate _________mg L^{-1}

   Site _______________

   Horizon _______________

   Depth_______________ cm

   Phosphate _________mg L^{-1}

   Nitrate _________mg L^{-1}

   Site _______________

   Horizon _______________

   Depth_______________ cm

   Phosphate _________mg L^{-1}

   Nitrate _________mg L^{-1}

Questions 8

Why does nutrient content only potentially determine soil fertility? _______________

____

How do your measurements of nutrient content relate to organic content of the soils you tested? _______________

____

SOIL TEXTURE

Weathering breaks rocks into progressively smaller particles. These particles consist of **minerals**, which are naturally occurring inorganic compounds usually made of two or more elements. All soils contain three kinds of **soil particles**: sand, silt, and clay (fig. 3.5). Gravel (> 2.0 mm) may be mixed with the soil. Clays are the final product of weathering and the smallest soil particles:

Type of Soil Particle	Diameter of Soil Particle (mm)
Sand	2–0.05
Silt	0.05–0.002
Clay	< 0.002

Different kinds of soils contain different proportions of sand, silt, and clay. Particle size is particularly important because it affects moisture content. Coarse-grained soils drain quickly—a good thing for some plant species and not for others. Fine-grained soils retain moisture far longer, but their large surface-to-volume ratio causes nutrients to adsorb to the particles' surfaces and thus lowers overall fertility. Clay soils contain more than 30% clay particles, and sandy soils contain less than 20% silt and clay. Soils containing approximately equal mixtures of sand, silt, and clay are called **loams**. Plants grow best in loams.

Procedure 3.8

Determine the particle size distribution of a soil sample.

1. Collect a soil sample ($\approx$ 75 g) from the horizons and sites selected by your instructor.
2. Air dry the sample in an open container for 24 h.
3. Gently use a mortar and pestle or rolling pin to break any clods.
4. Pass the soil through a 2-mm sieve to remove gravel and larger components. Weigh and record in step 20 the gravel retained by the sieve.
5. Oven dry (110°C, 24 h) at least 50 g of the soil that passed through the sieve.
6. Add 50 g of the sample to a 1-L beaker. Record 50 g in step 10 as the oven-dry weight of soil. Add 2 g of the detergent sodium hexametaphosphate and mix with the soil.
7. Add 500 mL distilled water to the beaker.
8. Use an electric mixer for 10 min to mix the suspension and disperse the soil particles. Mix for 5 more min if the soil is high in silt and clay.
9. Rinse ALL of the sample from the beaker, including any settled sand, into a 1000-mL graduated cylinder. Bring the total volume to 1000 mL by adding distilled water.
10. Cap the cylinder and invert the suspension several times. Avoid making suds at the surface.
11. Remove the cap and immediately add a soil hydrometer graduated in grams of soil per liter (g soil L^{-1}).
12. Record in step 20 the hydrometer value at the top of the meniscus 40 sec after placing it in sample. Measure and record the water's temperature.
13. Remove and clean the hydrometer. Leave the cylinder and suspension undisturbed for 6 h. After 6 h gently add the hydrometer. Record in step 20 the 6-h hydrometer reading. Measure and record the water's temperature.
14. Pass the entire suspension through a 0.053-mm sieve with a gentle rinse.
15. Transfer the retained sand to a pre-weighed, open container suitable for oven drying. Oven dry 24 h at 110°C. Weigh the oven-dried sand and record the value in step 20 as dry weight of sand.
16. Calculate and record in step 20 the % sand as:

    % sand = (dry weight of sand × 100) / (dry weight of soil)

17. Correct the 6-h hydrometer reading for temperature as:

    corrected 6-h hydr. reading = 6-h reading + 0.36 g L^{-1} for every 1° above 20°C

    (If temperature is below 20°C, subtract 0.36 g L^{-1} for every 1° below 20°C).

18. Calculate and record in step 20 the % clay as:

    % clay = (corrected 6-h hydr. reading × 100) / (dry weight of soil)

19. Calculate and record in step 20 the % silt as:

    % silt = 100 − (% sand + % clay)

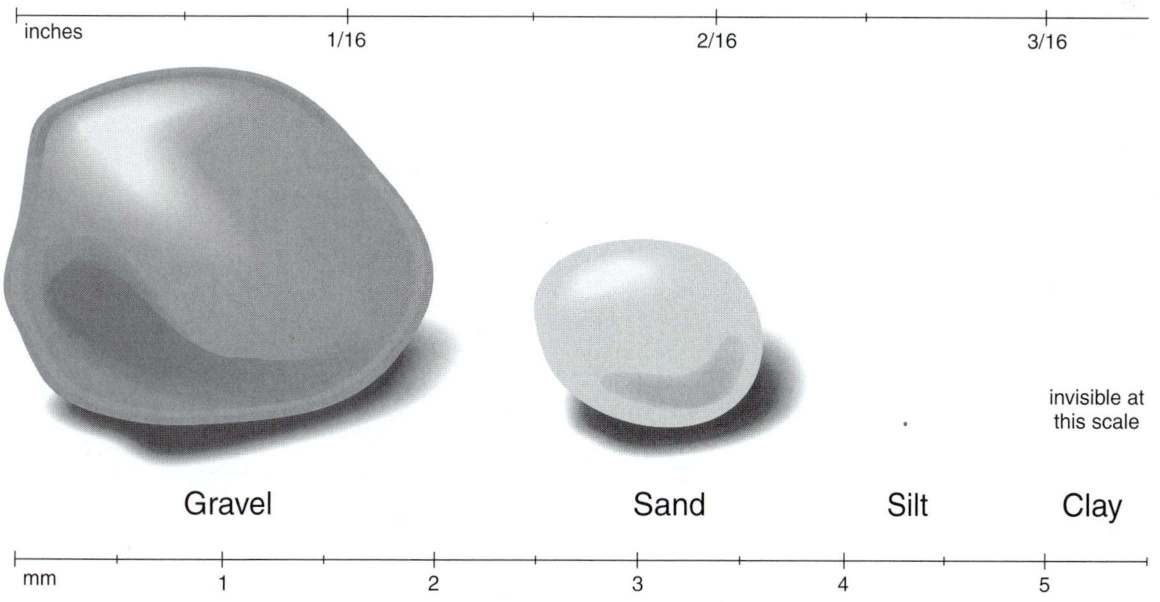

Figure 3.5

A comparison of gravel and soil particle sizes.

20. Site ______________________

Horizon ______________________

Depth ____________ cm

Sand ____________ %

Clay ____________ %

Silt ____________ %

Gravel ____________ g

Oven dry wt of soil ______ g

40-sec hydr. reading ______

40-sec temp __________ °C

6-h hydr. reading__________

6-h temp __________ °C

Oven drying container ___ g

Container with oven dried sand _____ g

Oven dried sand ________ g

Site ______________________

Horizon ______________________

Depth ____________ cm

Sand ____________ %

Clay ____________ %

Silt ____________ %

Gravel ____________ g

Oven dry wt of soil ______ g

40-sec hydr. reading ______

40-sec temp __________ °C

6-h hydr. reading__________

6-h temp __________ °C

Oven drying container ___ g

Container with oven dried sand _____ g

Oven dried sand ________ g

Site ______________________

Horizon ______________________

Depth ____________ cm

Sand ____________ %

Clay ____________ %

Silt ____________ %

Gravel ____________ g

Oven dry wt of soil ______ g

40-sec hydr. reading ______

40-sec temp __________ °C

6-h hydr. reading__________

6-h temp __________ °C

Oven drying container ___ g

Container with oven dried sand _____ g

Oven dried sand ________ g

Site ______________________

Horizon ______________________

Depth ____________ cm

Sand ____________ %

Clay ____________ %

Silt ____________ %

Gravel ____________ g

Oven dry wt of soil ______ g

40-sec hydr. reading ______

40-sec temp __________ °C

6-h hydr. reading__________

6-h temp __________ °C

Oven drying container ___ g

Container with oven dried sand _____ g

Oven dried sand ________ g

Questions 9

Does your temperature profile provide evidence for different particle size compositions having different temperature insulating properties? ______________________

____

Considering soil particle composition and moisture retention, why would it be somewhat misleading to state "Johnson grass grows best with rainfall of 30 in. per year"? _____

____

Questions for Further Thought and Study

1. Northern soils may have a *permafrost zone*. Research this term. What factors would affect the depth of this zone?

2. Particle size distribution affects nutrient availability. How would you design an experiment to determine the optimum particle distribution for fertile soil?

3. What are the most common soil amendments used by farmers? In other words, how do they manipulate their soil?

4. Speak to a local farmer. What are the best qualities and worst qualities of his soil?

Oxygen and Carbon Dioxide Cycling

4

Objectives

As you complete this lab exercise you will:

1. Detect the uptake of carbon dioxide during photosynthesis.
2. Measure the production of carbon dioxide by a variety of organisms during respiration.
3. Measure dissolved oxygen in a water sample.
4. Measure simultaneous changes in carbon dioxide and oxygen during community respiration and photosynthesis.
5. Measure the total oxygen demand to decompose a sewage effluent sample.

Organic molecules of the biosphere are rich with carbon-carbon bonds. These bonds hold the energy captured by photosynthesis using sunlight as energy and carbon dioxide as raw material. Each year the world's ecosystems use about 10% of the 700 billion metric tons of carbon dioxide to synthesize organic compounds. This carbon is constantly cycled among living organisms and between organisms and their environment. Respiration and photosynthesis drive that cycling (fig. 4.1).

Carbon and oxygen cycling are directly linked (fig. 4.2). Photosynthesis uses CO_2 to make organic molecules and liberates O_2. Respiration uses O_2 to accept electrons and oxidize carbon as it releases the energy of organic molecules and stores it in ATP. Respiration releases CO_2. The revolving uptake and production of oxygen and carbon dioxide are critical to the operation of ecosystems. This lab exercise presents techniques commonly used to detect and measure O_2 and CO_2 uptake and release.

UPTAKE OF CARBON DIOXIDE DURING PHOTOSYNTHESIS

pH is a measure of the acidic or basic properties of a solution; pH 7 is neutral. Solutions having a pH < 7 are acidic, and solutions having a pH > 7 are basic (fig. 4.3). In the following procedure you will use the pH indicator phenol red to detect uptake of CO_2 by a photosynthesizing aquatic plant, *Elodea*. Phenol red (phenol-sulfonphthalein) is a pH indicator that turns yellow in an acidic solution (pH < 7) and red in a neutral to basic solution (pH > 7).

To detect CO_2 uptake, you will put a plant into a solution that you have made slightly acidic with your breath. Carbon dioxide in your breath dissolves in water to form carbonic acid, which lowers the pH of the solution.

$$\underset{\text{water}}{H_2O} + \underset{\text{carbon dioxide}}{CO_2} \longleftrightarrow \underset{\text{carbonic acid}}{H_2CO_3} \longleftrightarrow \underset{\text{hydrogen ion}}{H^+} + \underset{\text{bicarbonate ion}}{HCO_3^-}$$

As a photosynthesizing plant fixes CO_2 and removes it from the solution, the pH rises. When the pH rises above 7, the dissolved indicator turns red.

Procedure 4.1

Detect the uptake of CO_2 during photosynthesis.

1. Fill two test tubes half full with a dilute solution of phenol red provided by your laboratory instructor.
2. Blow slowly into each solution with a straw. Stop blowing bubbles immediately when the color changes to yellow. Excess carbonic acid unnecessarily lengthens the procedure.
3. Add pieces of healthy *Elodea* totaling about 10 cm to a tube. Pour off excess solution above the *Elodea*.
4. Place both tubes in a bright, sunlit window or ≈ 0.5 m in front of a 100-watt bulb for 30–60 min.
5. Observe the tubes every 10 min and note any color change in the solution.

Questions 1

What happens to the color of the indicator? ____________

____

Why did the color change? ____________

____

How does this change in color relate to the summary equation for photosynthesis? ____________

____

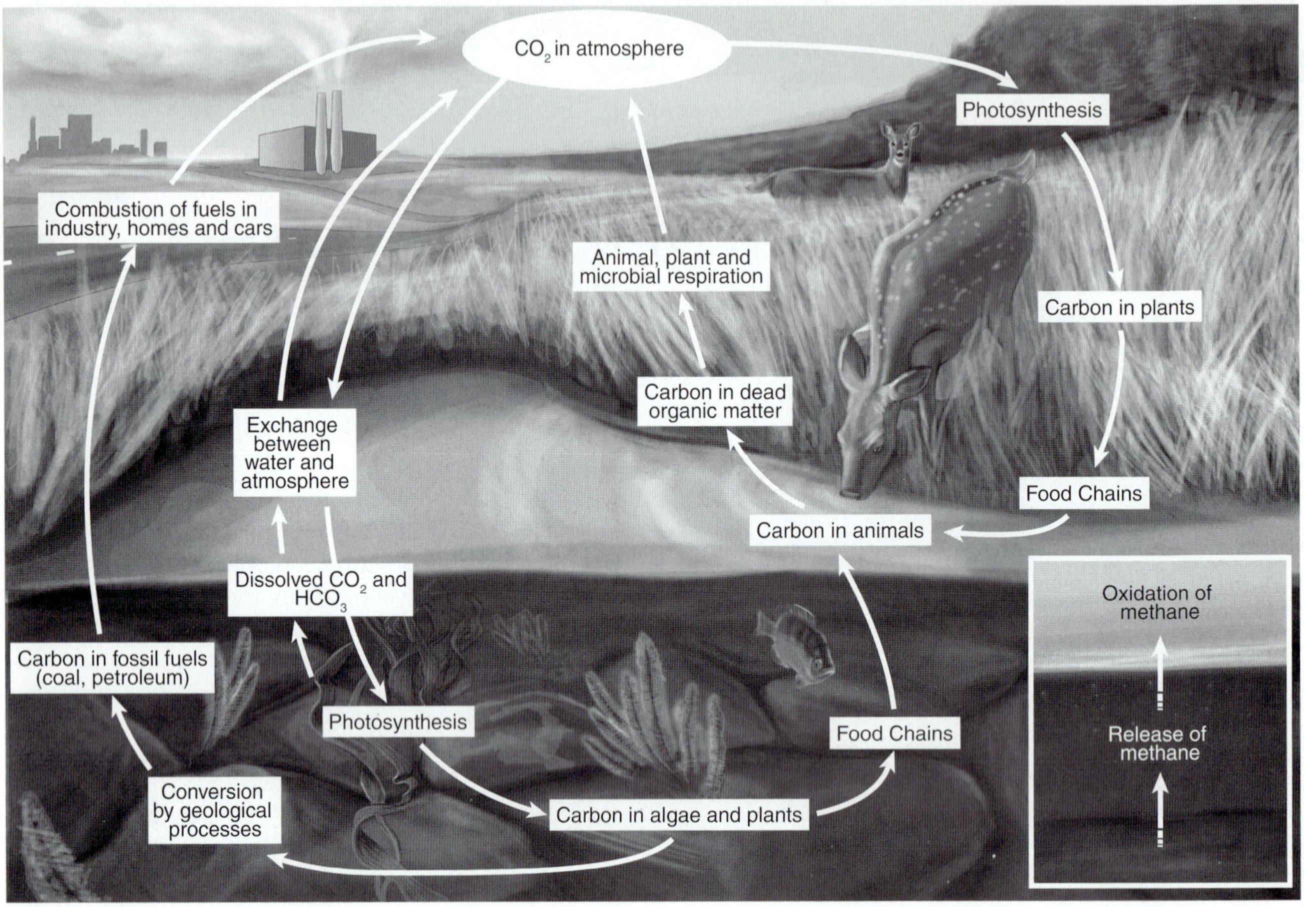

Figure 4.1

Photosynthesis by plants and algae captures carbon in the form of organic chemical compounds. Aerobic respiration by organisms and fuel combustion by humans return carbon to the form of carbon dioxide or bicarbonate. Microbial methanogens living in oxygen-free microhabitats, such as the mud at the bottom of the pond, might produce methane, a gas that would enter the atmosphere and then gradually be oxidized abiotically to carbon dioxide (shown in inset).

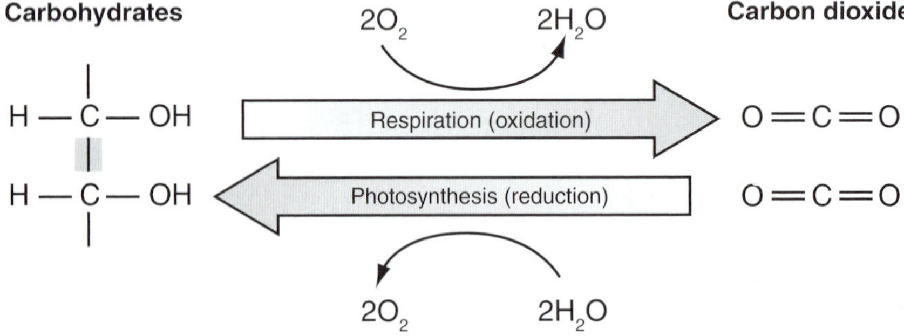

Figure 4.2

Carbon dioxide and oxygen cycling.

PRODUCTION OF CO_2 DURING AEROBIC RESPIRATION

Cellular respiration rates are important to ecologists because they represent the expenditure of energy by an organism. Active organisms with keen senses expend the majority of their assimilated energy by maintaining internal homeostasis and highly responsive nervous systems. More sedentary, tolerant, less-active organisms with a simple or absent nervous system require far less energy, and expend a smaller percent in respiration. Understanding respiration (and its release of CO_2) is part of understanding organism and community energetics.

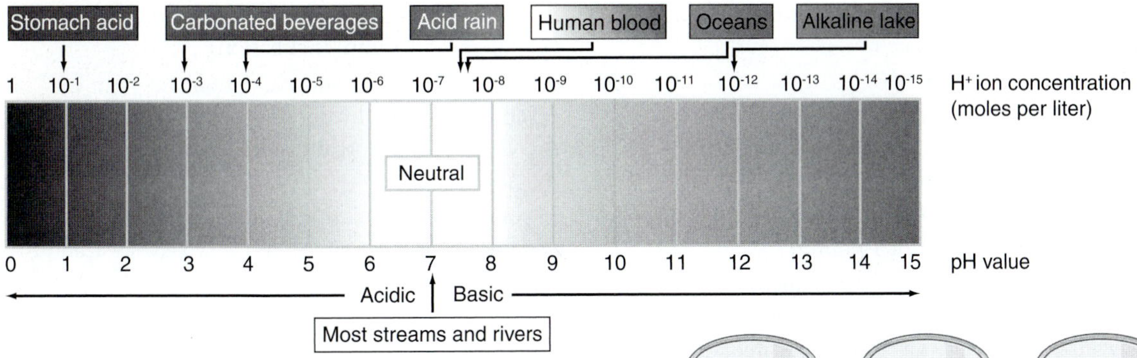

Figure 4.3

The pH scale of hydrogen ion concentration.

Procedure 4.2 uses the indicator phenolphthalein to detect changes in pH resulting from CO_2 and carbonic acid produced during cellular respiration (fig. 4.4). Phenolphthalein is red in basic solutions and colorless in acidic solutions. Thus, you can monitor cellular respiration as a change in pH due to carbonic acid production. In Procedure 4.2, you will directly measure the volume and milligrams of CO_2 produced by a respiring organism.

Questions 2

The organisms studied in Procedure 4.2 include a plant (*Elodea*) and an animal (snail). Which do you think will respire more? ________________

Write your hypothesis here: ________________

____

Procedure 4.2

Measure relative CO_2 production by aerobic organisms.

Experimental Setup

1. Obtain 400 mL of culture solution provided by your instructor. This solution has been dechlorinated and adjusted to be slightly acidic.
2. Place 100 mL of this solution in one labeled beaker for each organism tested (treatment beakers), plus one control beaker (table 4.1).
3. Obtain the organisms listed in table 4.1 from your instructor and determine the volume of each organism by following steps 4–6. Your instructor may include additional organisms and may ask you to work with mass rather than volume.

Determine Organism Volume by Water Displacement

4. Pour exactly 25 mL of water in a 50-mL graduated cylinder.
5. Measure the volume of each organism being tested. Place the organism in the cylinder and note the increase in volume above the original 25 mL. This increase equals the volume of the organism.

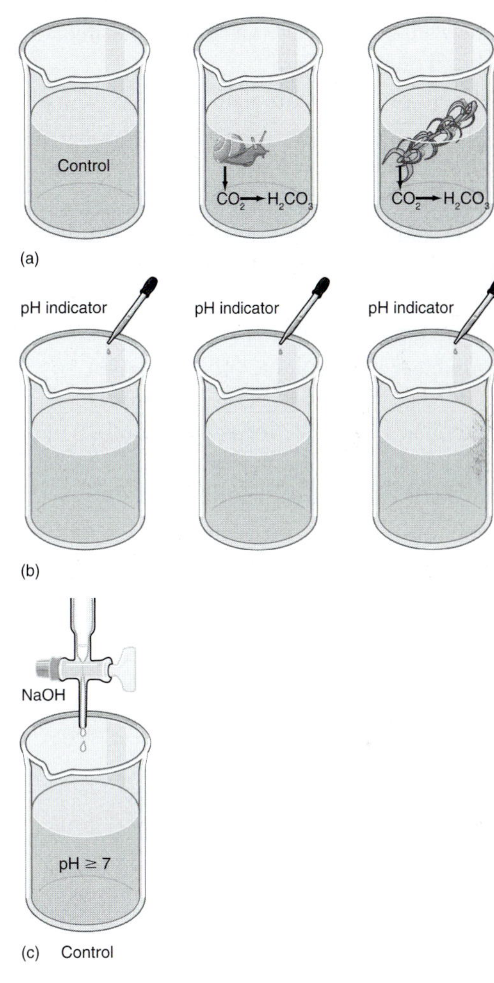

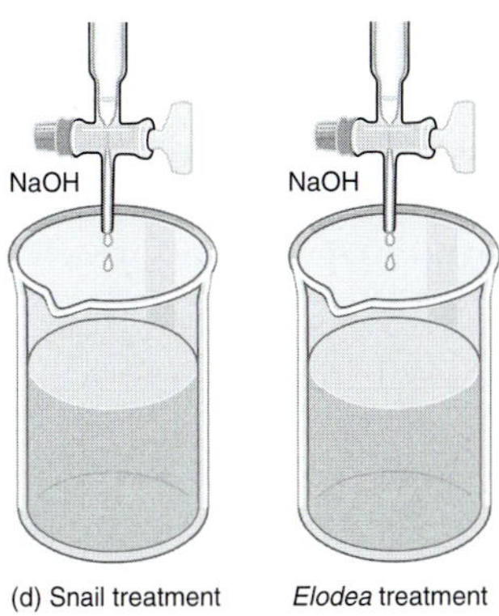

Figure 4.4

Illustration of the steps of Procedure 4.2.

6. Record the volumes in table 4.1. Gently return each organism to the appropriate beaker.

Incubate Experimental Treatments

7. Cover each beaker with a plastic film or petri dish top and set them aside on your lab bench. Place the beaker containing the *Elodea* in the dark by covering it with a coffee can or aluminum foil.
8. Allow the organisms to respire for 15 min.
9. Gently remove the organisms from the treatment beakers and return them to their original culture bowls. Disturb the water in the beakers as little as possible.

Titrate to Gather Your Raw Data

10. Add four drops of phenolphthalein to the contents of each beaker. The solutions should remain clear because the solutions are acidic.
11. Obtain a burette or dropper bottle to dispense NaOH (0.0227 N). Add NaOH drop-wise to the contents of the control beaker. Swirl the contents of the beaker after each drop. An accurate endpoint remains pink for only a moment. Do not over-titrate.
12. Record in table 4.1 the volume (mL) of NaOH required to reach the control beaker endpoint of phenolphthalein (i.e., make the solution pink). Either read the volume directly from the burette, or calculate volume as 20 drops per milliliter.
13. Repeat steps 11 and 12 for the treatment beakers. Record in table 4.1 the titration volume of NaOH to reach endpoint for each treatment beaker.

Calculate Your Results

14. For each treatment subtract the volume of NaOH to reach endpoint for the control beaker from the volume of NaOH added to each treatment beaker and record these values in table 4.1.
15. If 100 mL was the original beaker volume, then the *milliliters* of titrant (mL 0.0227 N NaOH) to reach endpoint (calculated in step 14) equal the *milligrams* CO_2 per liter in the solution. Record these equivalent values in table 4.1 as concentration of CO_2 produced by respiration.
16. Calculate the milligrams of CO_2 produced per hour by respiration by dividing the concentration per liter by 10 (to adjust for 100 mL volume) and then by 4 (to adjust for 15-min. incubation).
17. Calculate the milligrams CO_2 produced per mL organism per hour by dividing mg CO_2 produced by the volume of the organisms.
18. Record in table 4.2 the data from all student groups in your class. Calculate the mean, standard deviation, and 95% confidence interval (see Exercises 1 and 2) for each of the two tested organisms.

Questions 3

Do the 95% confidence intervals surrounding the means of CO_2 production by the two tested organisms overlap?

____

Before you gathered your data, you formulated a hypothesis about the expected results. Do you reject or do you fail to reject your hypothesis? _______________

The last two columns in table 4.1 express results differently. Which is the most appropriate for assessing the CO_2 production of an organism? _______________

____

What is your major conclusion from this procedure?

____

____

What characteristics of the organisms most likely contributed to their differences in CO_2 production rate? _______

____

____

Are your conclusions applicable to all plants and animals, or only to the organisms you tested? _______________

____

What other organisms would you include in an expanded experiment? Why did you choose these organisms? _____

____

____

Table 4.1

Measurement of CO_2 production during respiration

Organisms	Total Volume of Organisms (mL)	Volume NaOH to Reach Endpoint	Volume NaOH to Reach Endpoint Minus Control Volume	Concentration of CO_2 Produced by Respiration	CO_2 Produced by Respiration per Hour	CO_2 Produced mL^{-1} Organism h^{-1}
Beaker 1: Four snails	_____ mL	_____ mL	_____ mL	___ mg CO_2 L^{-1}	___mg CO_2 h^{-1}	___ mg CO_2 mL^{-1} organism h^{-1}
Beaker 2: *Elodea*	_____ mL	_____ mL	_____ mL	___ mg CO_2 L^{-1}	___mg CO_2 h^{-1}	___ mg CO_2 mL^{-1} organism h^{-1}
Beaker 3: _____	_____ mL	_____ mL	_____ mL	___ mg CO_2 L^{-1}	___mg CO_2 h^{-1}	___ mg CO_2 mL^{-1} organism h^{-1}
Control beaker		_____ mL				

TABLE 4.2

	CO_2 produced mL^{-1} *Elodea* h^{-1}	___ mg CO_2 mL^{-1} snail h^{-1}
Student group 1		
Student group 2		
Student group 3		
Student group 4		
Student group 5		
Mean =		
Std. dev. =		
Confidence interval =		

OXYGEN AND CARBON DIOXIDE CYCLING VIA PHOTOSYNTHESIS AND RESPIRATION

Aquatic samples are ideal for measuring gas exchange because dissolved gases are handled more easily in the lab than are atmospheric gases. Furthermore, a plankton community and its microorganisms in lake water is a convenient microcosm of both autotrophs and heterotrophs.

The ecological significance of dissolved gases in aquatic systems is further magnified by the fact that gasses, especially oxygen, do not dissolve in high concentrations in water, and warmer water holds less oxygen than does cold water. Only 8 mg oxygen L^{-1} will dissolve in water at 25°C. This translates into 8 parts per million and is far less than the 200,000 ppm of oxygen in the atmosphere. Cold water (4°C) can hold (i.e., is saturated by) 12.5 mg oxygen L^{-1}. In the next three procedures you will (1) refine your technique of measuring dissolved oxygen; (2) learn to use a water sampler to gather lake water; and (3) measure the cycling of O_2 and CO_2 within a plankton community.

A common method used to measure dissolved oxygen (DO) in a water sample is the **Winkler titration method**. Although specialized electronic meters can measure DO, the Winkler titration is simple and accurate. Measuring DO is an important part of procedures to detect oxygen production and uptake.

Procedure 4.3

Practice the Winkler titration method for measuring dissolved oxygen.

1. Examine a 300-mL DO bottle (fig. 4.5). Its narrow neck minimizes the water's contact with air. Its ground-glass stopper is precisely shaped to enclose exactly 300 mL.
2. Fill a 1-L flask with tap water, and swirl it vigorously to aerate the water.

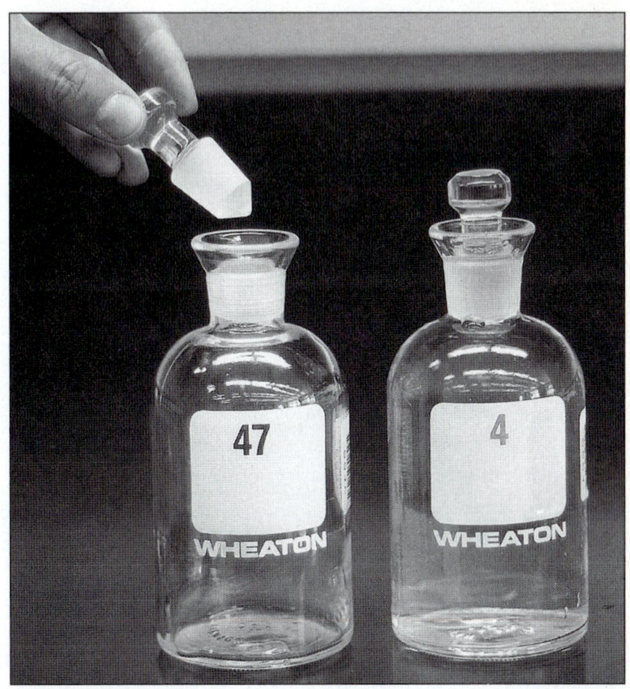

Figure 4.5

Dissolved oxygen bottles are designed to hold a precise amount of water, minimize contact with air, and be chemically inert. Each bottle has an identification number and a ground glass stopper that seals the water from air when properly inserted.

3. Fill a clean DO bottle from the flask and stopper the bottle so no bubbles are trapped. Record the ID number of this DO bottle for aerated water in table 4.3.
4. Boil the remaining water in the flask for 5 min. This reduces the dissolved gases.
5. Let the boiled water cool for 10–15 min and then fill a second DO bottle. Fill the bottle gently to minimize mixing with air. Record the ID number of the DO bottle in table 4.3.
6. Assemble the materials needed for a Winkler titration.
7. Follow the steps listed in the boxed insert "Winkler Titration: Chemistry and Procedure" to determine the DO concentrations in the two water samples. The volume of each DO bottle allows three replicate 100-mL titrations.
8. Record your results in table 4.3.

Question 4

Examine the data in table 4.3. How did boiling affect the dissolved oxygen concentration of the water sample?

___

___

Winkler Titration: Chemistry and Procedure

Materials:

Manganese sulfate solution	Sodium thiosulfate, or PAO titrant
Alkali-iodine-azide solution	Burette, 20-mL
Sulfuric acid	Ring stand
Starch indicator solution	250-mL flask
Volumetric pipets and bulbs	

The Winkler titration procedure for measuring dissolved oxygen begins with collecting a water sample and adding a series of chemicals (manganese sulfate, alkali-iodine-azide, and sulfuric acid). The first chemical reaction forms a precipitate of manganous hydroxide.

$$MnSO_4 + 2\ KOH \rightarrow Mn(OH)_2\downarrow + K_2SO_4$$

As this precipitate settles through the water sample, it quickly absorbs dissolved oxygen molecules present according to the following equation:

$$2\ Mn(OH)_2 + O_2 \rightarrow 2\ MnO(OH)_2$$

Adding iodide and then acid releases iodine (I_2) in amounts equivalent to the O_2 that was originally present.

$$MnO(OH)_2 + 2\ KI + H_2O \rightarrow Mn(OH)_2 + I_2 + 2\ KOH$$

A starch indicator is added. The iodine (I_2) reversibly binds with the starch and makes the solution dark blue-black. The iodine is quantitatively removed by slowly titrating with (adding) a standardized solution of sodium thiosulfate or PAO solution until the dark color of the solution disappears (titration endpoint). The amount of titrant used to remove the blue-black color is directly proportional to the amount of I_2, which is directly proportional to the amount of O_2 originally dissolved in the water sample.

STEPS FOR SAMPLE FIXATION

1. Obtain a water sample in a 300-mL, glass-stoppered DO bottle.
2. Remove the DO bottle's stopper and insert the tip of a pipet just below the water level, and add 2 mL $MnSO_4$ solution.
3. Insert the tip of a pipet just below the water level, and add 2 mL alkaline-iodine-azide solution.
4. Replace the stopper and invert the bottle 10 times to mix the solution.
5. Allow the resulting precipitate to settle until the top third of the sample is clear. Then invert the bottle 10 times again to remix the solution.
6. Allow the resulting precipitate to settle until the top third of the sample is clear.
7. CAUTION Remove the stopper and add 2 mL concentrated H_2SO_4 by releasing the acid from a pipet tip held against the inside of the upper lip of the sample bottle. Allow the acid to flow down the neck of the bottle into the water sample.
8. Replace the stopper and invert the bottle 10 times to mix the solution. The sample is now "fixed" and can be stored for 3 days with refrigeration before titrating.

STEPS FOR SAMPLE TITRATION

1. Obtain the fixed sample to be titrated.
2. Measure 100 mL of solution with a graduated cylinder and pour it into a 250-mL flask.
3. Add 1 mL of starch indicator solution.
4. Slowly titrate drop-wise with 0.0125 N thiosulfate solution (or PAO) until the blue color first disappears. Disregard any subsequent reappearance of the blue color.
5. Record the milliliters of titrant used.
6. One milliliter of 0.0125 N titrant equates to a 1 mg of DO in a liter of sample. Convert the milliliters of titrant to mg L^{-1} of DO. For example, 7.5 mL of titrant equates to 7.5 mg DO L^{-1} in lake water.

TABLE 4.3

TITRATION VOLUMES AND DISSOLVED OXYGEN CONCENTRATIONS FOR AERATED AND BOILED WATER SAMPLES.

	DO Bottle ID Number ____ Aerated Water Sample		DO Bottle ID Number ____ Boiled Water Sample	
	mL of titrant	DO (mg L^{-1})	mL of titrant	DO (mg L^{-1})
Replicate 1	_______	= _______	_______	= _______
Replicate 2	_______	= _______	_______	= _______
Replicate 3	_______	= _______	_______	= _______
	$\bar{x}$ =	_______	$\bar{x}$ =	_______

4–6

Procedure 4.4

Collect an undisturbed lake-water sample for dissolved oxygen analysis.

1. Obtain a 300-mL DO bottle and ground-glass stopper. Rinse the inside of the bottle with some of the lake water being sampled.
2. Examine a van Dorn water sampler. Close the drain valve at the base of the flexible drain tube (fig. 4.6).
3. Your instructor will show you how to CAREFULLY set the spring-loaded suction cups, and how to trigger the release mechanism with a weighted messenger.
4. Cock the van Dorn sampler. Slowly submerge the cocked van Dorn completely in lake water (or a sink full of water if you are practicing). Lower the van Dorn to 1-m depth as marked on the line.
5. Move the van Dorn sampler about 1 m horizontally to displace any previously disturbed water in the cylinder.
6. Drop the messenger to close the cylinder and enclose the water sample.
7. Raise the cylinder out of the water and rest it vertically on the edge of a solid surface so the drain valve is at the lower end.
8. Insert the drain tube into the DO bottle so the end of the tube touches the bottom of the bottle.
9. Open the drain valve. If water doesn't flow freely into the bottle, lift the edge of the upper suction cup to break the seal and allow air flow.
10. Allow the water to overflow the DO bottle until the volume of the bottle has been displaced three times.
11. As the water continues to flow, slowly pull the tube out of the bottle.

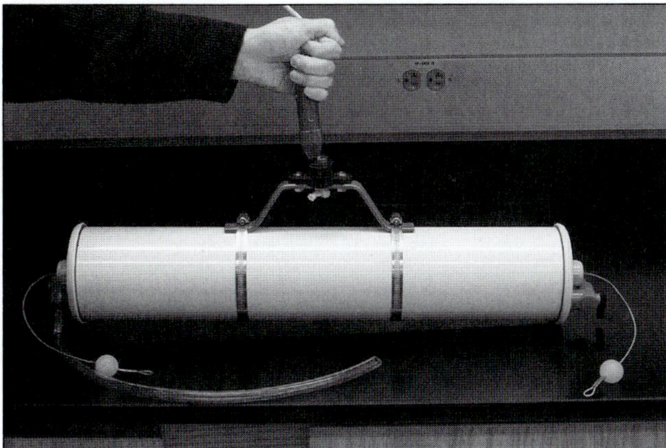

Figure 4.6

A van Dorn water sampler effectively captures a water sample from a known depth when a heavy "messenger" travels down the rope to trigger rubber cups and seal water inside a large cylinder. The water is brought to the surface and drained into a sample bottle by opening a valve and hose on the side of the sampler.

12. Insert the ground-glass stopper into the bottle to seal the 300-mL volume with no bubbles. The sample is now ready for Winkler titration.
13. Empty the van Dorn sampler back into the lake.

Question 5

Why should the volume be overflowed three times while filling a DO bottle? ______________________

____

Procedure 4.5

Use light and dark DO bottles to measure simultaneous photosynthesis and respiration rates of a plankton community.

1. Discuss with your instructor where to sample a nearby lake with a rich plankton community. Your instructor may assign measurements for a different depth to each group in the class.
2. Assemble six 300-mL DO bottles. Cover two of the bottles (= DO_{dark} bottles) completely (light-tight) with tinfoil (fig. 4.7).
3. Use a van Dorn sampler to take water samples 0.25-m deep according to Procedure 4.4, and fill the four light bottles and the two dark bottles with lake water. Use a small square of tinfoil to cover the stopper and neck of each dark bottle.

Figure 4.7

These light and dark bottles contain lake water with plankton. During incubation, photosynthesis and respiration change the dissolved oxygen in the light bottle. Respiration decreases the dissolved oxygen change in the dark bottle. The rate of photosynthesis is calculated by comparing changes in dissolved oxygen in each bottle. A bottle is made "dark" by tinfoil or a black plastic coating.

4. Designate two of the light bottles as DO_{init} bottles for initial dissolved oxygen measurement, and record their ID numbers in table 4.4.
5. Designate two of the light bottles as DO_{light} bottles. They will allow light and photosynthesis during incubation. Record their ID numbers in table 4.4.
6. Designate two of the dark bottles as DO_{dark} bottles. They will not allow photosynthesis during incubation. Record their ID numbers in table 4.4.
7. Suspend the four bottles (two light, two dark) 0.25 m beneath the surface from a flotation device to incubate for 24 h. Your instructor may modify the incubation time.
8. Immediately after suspending the light and dark bottles, fix the contents of the DO_{init} bottles according to the STEPS FOR SAMPLE FIXATION in the boxed insert: Winkler Titration Chemistry and Procedure.
9. Return to the lab and titrate the two DO_{init} bottles according to the STEPS FOR SAMPLE TITRATION in the boxed insert: Winkler Titration: Chemistry and Procedure.
10. Record the titration results for the three 100-mL aliquots from each bottle in table 4.4 as mL of titrant.
11. After 24 h incubation, retrieve the DO_{light} and DO_{dark} bottles. While in the field, fix their contents according to the STEPS FOR SAMPLE FIXATION.
12. Return to the lab and titrate the samples according to the STEPS FOR SAMPLE TITRATION.
13. Record the titration results for the three 100-mL aliquots from each bottle in table 4.4 as mL of titrant.
14. Calculate and record in table 4.4 the mean milliliters of titrant for each set of six values.
15. The mean milliliters of titrant equals the mg L^{-1} DO for each treatment. Record these values in table 4.4.
16. Calculate and record community respiration in table 4.4.

    community respiration $= DO_{init} - DO_{dark}$
17. Calculate and record net and gross photosynthesis in table 4.4.

    net photosynthesis $= DO_{light} - DO_{init}$

    gross photosynthesis = community respiration + net photosynthesis
18. Be prepared to combine the data for your assigned depth with data of the other groups for an analysis of the production-depth profile.

Questions 6

If your class measured photosynthesis and respiration at different depths, does the productivity rate reflect the depth of the water sampled? ____________________

____

If the biomass and number of planktonic animals far exceed the amount of algae, how would that affect each of the three community variables calculated in steps 16 and 17?

____

____

TABLE 4.4

DATA AND CALCULATIONS FOR MEASUREMENT OF PHOTOSYNTHESIS AND RESPIRATION IN LAKE WATER

	Bottle ID Number	mL Titrant for 100-mL Aliquot			Dissolved Oxygen
DO_{init} bottle	_____	_____ mL	_____ mL	_____ mL	
DO_{init} bottle	_____	_____ mL	_____ mL	_____ mL	
				$\bar{x}$ = _____	= _____ mg L^{-1} DO = DO_{init}
DO_{dark} bottle	_____	_____ mL	_____ mL	_____ mL	
DO_{dark} bottle	_____	_____ mL	_____ mL	_____ mL	
				$\bar{x}$ = _____	= _____ mg L^{-1} DO = DO_{dark}
DO_{light} bottle	_____	_____ mL	_____ mL	_____ mL	
DO_{light} bottle	_____	_____ mL	_____ mL	_____ mL	
				$\bar{x}$ = _____	= _____ mg L^{-1} DO = DO_{light}

Community respiration	Net photosynthesis	Gross photosynthesis
_____ mg O_2 $L^{-1}d^{-1}$	_____ mg O_2 $L^{-1}d^{-1}$	_____ mg O_2 $L^{-1}d^{-1}$

4–8

BIOCHEMICAL OXYGEN DEMAND

Not all communities are based on autotrophic plants. Some communities with high inputs of organic matter and little light for photosynthesis are **heterotrophic** and not driven by photosynthesis. They often experience critical oxygen deficits from decomposition of their high organic content. Microbial communities such as lake sediments, leaf litter, sewage effluent, and polluted lakes and rivers are typically heterotrophic, and their O_2 deficits and high organic contents dictate community structure. Their organic load may be so significant that ecologists sometimes characterize it in terms of the amount of O_2 required for its decomposition—sometimes called **biochemical oxygen demand (BOD)**.

The organic content of heterotrophic communities and its impact on CO_2 and O_2 cycling can be bioassayed by measuring BOD. BOD is the amount of oxygen required by aerobic microorganisms to decompose the organic matter in a sample of water. BOD is a common measure of organic pollution in a water sample or in a diluted sample of organic soil or sediment. A BOD assay measures the dissolved oxygen consumed as microbes respire and break down organic matter in the sample.

A BOD procedure includes collecting water samples of a heterotrophic community, measuring the initial oxygen content, incubating the samples for 5 days, and measuring the final DO concentration. The difference in the initial DO and the DO after incubation is the biochemical oxygen demand. Most pristine rivers have a 5-day incubation BOD of less than 1 mg O_2 L^{-1}. Moderately polluted rivers have BODs from 2–8 mg O_2 L^{-1}. Wastewater coming into most sewage treatment plants is about 200 mg L^{-1}. Efficiently treated sewage has a BOD value of about 20 mg L^{-1}.

Procedure 4.6

Measure the biochemical oxygen demand in a highly organic water sample.

1. Assemble eight numbered, 300-mL DO bottles with stoppers, and either graduated or volumetric pipets to measure 20–50 mL volumes, and materials needed for the Winkler oxygen method.
2. Fill a large (> 2 L) flask with demineralized water, and shake it vigorously to saturate it with dissolved oxygen.
3. Collect a suitable sample of highly organic sewage effluent. Your instructor will describe the risks and proper procedures for handing samples of effluent.
4. To avoid the dissolved oxygen being completely depleted during incubation, the sample should be diluted. To do this, add 150 mL of sample to two DO bottles (0.5 dilution factor). Label the sample bottles DILUTE0.50.
5. Add 30 mL of sample to two DO bottles (1.0 dilution factor). Label the sample bottles DILUTE0.10.
6. Add 15 mL to two DO bottles (0.05 dilution factor). Label the sample bottles DILUTE0.05. Record the bottle ID numbers in table 4.5. Your instructor may recommend a different dilution scheme.
7. Fill the remaining volume of all DO bottles (including the two control bottles) with aerated, demineralized water from the flask to slightly above the neck of each DO bottle and stopper them to capture 300 mL with no air bubbles.
8. Use Winkler titration to determine the initial dissolved oxygen concentration for one bottle of each of the four pairs of samples including the controls. Record these initial values in table 4.5.
9. Incubate the remaining four samples at 20°C for 5 days.
10. After incubation, use Winkler titration to determine the dissolved oxygen concentration for each of the four samples including the control. Record these incubation values in table 4.5.
11. For each sample bottle, average the three 100-mL replicate titrations, and record the mean in table 4.5.
12. Remember that the milliliters of titrant equals the milligrams O_2 per liter (mg O_2 L^{-1}) of the sample. Calculate and record the control adjustment as:

    control adjustment = $\bar{x}$ initial − $\bar{x}$ incubated = ________ mg L^{-1}

13. Calculate and record the changes (Δ) in DO after incubation for the three dilutions as:

    For dilution 0.50,
    $\Delta DO = DILUTE0.50_{initial} - DILUTE0.50_{incubated} - \text{control adjustment}$

    For dilution 0.10,
    $\Delta DO = DILUTE0.10_{initial} - DILUTE0.10_{incubated} - \text{control adjustment}$

    For dilution 0.05,
    $\Delta DO = DILUTE0.05_{initial} - DILUTE0.05_{incubated} - \text{control adjustment}$

14. Discard data for any samples in which the $DILUTE_{incubated}$ value was 0.0. These treatments became anoxic (DO = 0.0 mg L^{-1}) during incubation and are inaccurate.
15. Select the dilution that produced a DO drop (ΔDO) of 2–4 mg L^{-1} from its initial value. For this dilution, calculate and record the final BOD value in table 4.5 as:

    $BOD = (DO_{init} - DO_{incubated}) / \text{dilution factor}$

Questions 7

Does your effluent sample have a BOD similar to the 20 mg L^{-1} typical for other treated sewage samples?____________

____

Oxygen and carbon dioxide cycling is critical to healthy communities. Which of the two gases is likely the limiting factor for organically polluted systems? _______________

____

Are heterotrophic communities naturally self-sustaining? How so? ______________________________

____

TABLE 4.5

DATA AND CALCULATIONS FOR MEASUREMENT OF BIOCHEMICAL OXYGEN DEMAND

	Bottle ID Number	mL titrant for 100-mL aliquot				Δ DO
$CONTROL_{init}$	_____	____ mL	____ mL	____ mL	$\bar{x}$ initial = ____ mL	
$CONTROL_{incub}$	_____	____ mL	____ mL	____ mL	$\bar{x}$ incubated = ____ mL	Control adjustment = _______ mg L^{-1}
				control adjustment = $\bar{x}$ initial mL − $\bar{x}$ incubated mL		
$DILUTE0.50_{init}$	_____	____ mL	____ mL	____ mL	$\bar{x}$ initial = ____ mL	
$DILUTE0.50_{incub}$	_____	____ mL	____ mL	____ mL	$\bar{x}$ incubated = ____ mL	**Dilution 0.50**
			Δ DO = $DILUTE0.50_{initial}$ − $DILUTE0.50_{incubated}$ − control adjustment			Δ DO = ___ mg L^{-1}
$DILUTE0.10_{init}$	_____	____ mL	____ mL	____ mL	$\bar{x}$ initial = ____ mL	
$DILUTE0.10_{incub}$	_____	____ mL	____ mL	____ mL	$\bar{x}$ incubated = ____ mL	**Dilution 0.10**
			Δ DO = $DILUTE0.10_{initial}$ − $DILUTE0.10_{incubated}$ − control adjustment			Δ DO = ___ mg L^{-1}
$DILUTE0.05_{init}$	_____	____ mL	____ mL	____ mL	$\bar{x}$ initial = ____ mL	
$DILUTE0.05_{incub}$	_____	____ mL	____ mL	____ mL	$\bar{x}$ incubated = ____ mL	**Dilution 0.05**
			Δ DO = $DILUTE0.05_{initial}$ − $DILUTE0.05_{incubated}$ − control adjustment			Δ DO = ___ mg L^{-1}
					BOD = (Δ DO)/dilution factor	**BOD = ____ mg L^{-1}**

Questions for Further Thought and Study

1. In what environments would oxygen cycling be unimportant?

2. How would you adapt the BOD procedure to determine the oxygen demand of solid sludge?

3. A surprising number of coral species and other reef invertebrates have algae growing symbiotically in their tissues. What are some adaptive advantages to that relationship?

4. Oxygen dissolves more readily in cold water than in warm water. Yet, deep lake water is often oxygen poor. Why is this the case?

Population Growth

5

Objectives

As you complete this lab exercise you will:

1. Describe how populations grow.
2. Show the effects of resources and environmental conditions on population growth.
3. Graphically analyze how the human population is increasing.

A population can grow incredibly fast. Abundant resources of food, space, and nutrients usually produce unlimited growth and a population growth curve with a geometric or "J" shape (fig. 5.1). Life history traits that influence the speed of population growth include (1) survival rates through reproductive age; (2) the age of first reproduction; (3) the number of offspring per generation; and (4) the number of times the species reproduces in a lifetime.

The simplest population model is one based on unlimited growth occurring in discrete pulses. An annual plant that reproduces once per year, or a protozoan that reproduces asexually once every hour by dividing to make two, is growing in discrete pulses according to the **geometric population growth** model. We compute this growth as:

$$N_t = N_0 \lambda^t$$

where

N_t = the number at time t
N_0 = the number of individuals present initially
λ = average number of offspring left by an individual during one time interval
t = number of time intervals (generations)

This calculation shows how quickly populations can grow, and the result is shown in figure 5.1. Consider *Escherichia coli*, a common bacterium that divides every 20 min in ideal conditions. In one day (1440 min), these bacteria can go through 72 (1440/20) generations. Therefore, if we start our experiment with one bacterium, the number of bacteria present after one day would be:

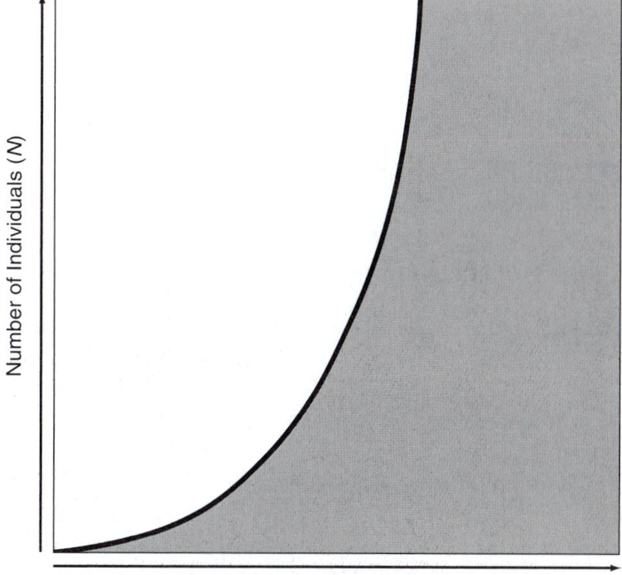

Figure 5.1
A J-shaped, geometric curve.

= $(1)(2^{72})$ bacteria
= 40,000,000,000,000,000,000,000 bacteria
= 4.7×10^{21} bacteria

This many bacteria would weigh an incredible 2.3 million kg (520 tons). But bacteria aren't the only organisms with high biotic potential. For example,

- Oysters each produce about 50 million eggs per year.
- A single pair of Atlantic cod and their descendants reproducing without hindrance would completely fill the Atlantic Ocean in 6 years.
- The 80 offspring produced every 6 months by a pair of cockroaches produce 130,000 roaches in only 18 months—enough to overrun any apartment!

Natural populations can grow at extraordinary rates dictated by high λ values and short generation times, but only for short periods and with unlimited resources.

Question 1

In simple terms, why isn't our world overrun with roaches and Atlantic cod if they can reproduce so dramatically?

____

____

ENVIRONMENTAL RESISTANCE AND CARRYING CAPACITY

Organisms in the "real world" do not usually reproduce at their maximum rates. Maximum rates are not sustainable because **environmental resistance** increases due to disease, accumulation of waste products, lack of food, and other factors. Ultimately, the size and growth of a population is a function of the environment as well as reproductive traits.

Question 2

Environmental resistance slows growth. Which of the four life-history traits listed at the beginning of this exercise would be influenced the most by environmental resistance?

____

____

To understand the effect of environmental resistance on growth rates, complete table 5.1. This table provides real data for a growing but limited population of bacteria. For comparison, you must calculate the size of a theoretical population of *E. coli* given unlimited resources. The population potentially doubles every 20 min. After doing these calculations, plot the growth of the theoretical and actual populations on figure 5.2.

Questions 3

How did growth of the actual population compare with that of the theoretical population during early stages of the experiment? During later stages? ______________________

____

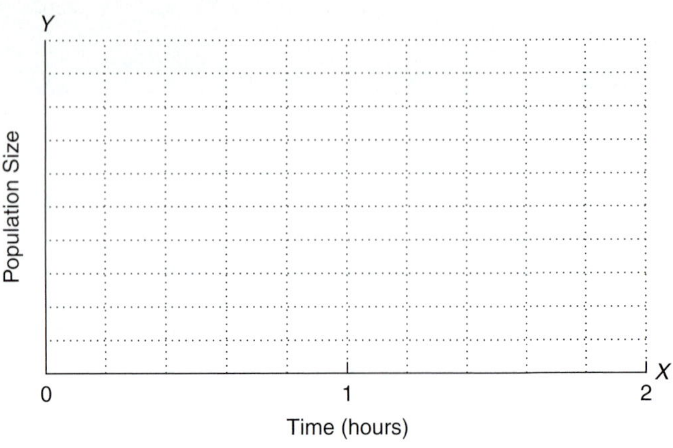

Figure 5.2

Theoretical and actual population growth of *E. coli*.

How long did it take the actual population to double during early stages of the experiment? Middle stages? Later stages?

____

____

When was growth of the actual population most rapid?

____

____

At what stage was growth slowest? What factors likely limited the growth? ______________________

____

As population size goes up, the environmental resistance increasingly slows the growth rate until it reaches zero and the population size remains constant. This is called **logistic population growth**, and the model for this is a sigmoid growth curve (fig. 5.3). Sustainable growth of a population occurs when the birth rate equals the death rate. This population size is referred to as the **carrying capacity** of the

TABLE 5.1

THEORETICAL AND ACTUAL GROWTH OF *E. COLI* BACTERIA

	Time		Size of Population (10^3 bacteria per mL)	
Generation	Hours	Minutes	Theoretical	Actual
1	0	0	8	8
2	0	20	16	15
3	0	40	32	28
4	1	0	______	48
5	1	20	______	120
6	1	40	______	220
7	2	0	______	221

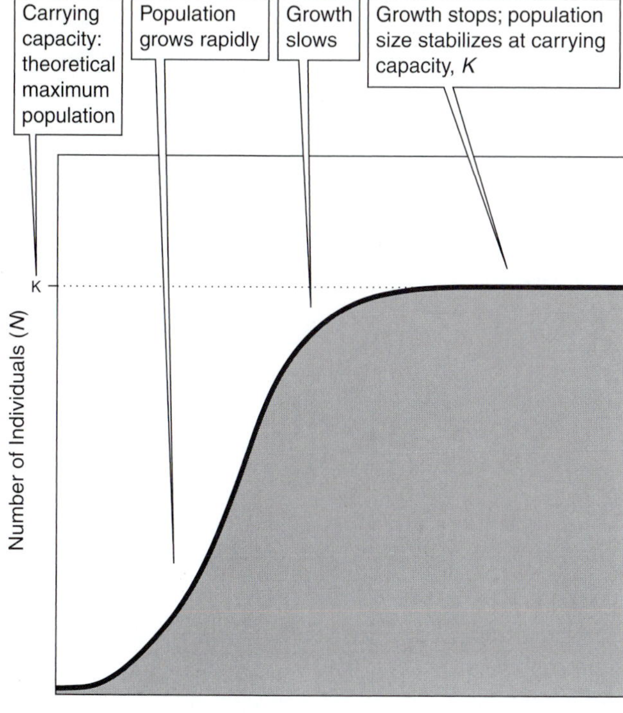

Figure 5.3

The theoretical sigmoid curve of population growth. The early lag and log phases closely represent geometric growth before environmental resistance and limited resources become significant.

environment. Population size remains near the carrying capacity as long as limiting factors are constant. However, that is rarely the case, and oscillations (occasional peaks and crashes) typically occur, especially for populations regulated primarily by abiotic factors.

In the laboratory, you can measure the growth of real populations such as bacteria that reproduce quickly. As bacteria reproduce in a clear nutrient broth, the broth becomes turbid. You can't accurately count individual bacteria in this broth, but you can measure the increase in turbidity of a growing culture. More turbidity means more bacteria—turbidity values roughly estimate population size.

Your instructor previously inoculated some test tubes of culture media with *E. coli*, a common bacterium. At regular time intervals, some of the tubes were put into a refrigerator to stop growth. Examine the cultures according to Procedure 5.1.

Procedure 5.1

Measure population growth of bacteria.

1. Discuss with your instructor proper technique for safely handling bacteria.
2. Examine cultures of *E. coli* grown for 0, 4, 8, 12, 24, and 48 h.

Table 5.2

Growth of Bacteria in a Limited-Nutrient Medium

Time (hours)	Turbidity Intensity (0–10)	Absorbance Value
0	______	______
4	______	______
8	______	______
12	______	______
24	______	______
48	______	______

3. Visually quantify the relative turbidity of each culture between 0 (clear) and 10 (most turbid).
4. Record your results in table 5.2.
5. If turbidometers are available, measure the turbidity of the solutions according to procedures demonstrated by your instructor. Spectrophotometers may also be used at 600 nm. Record your results in table 5.2.

Procedure 5.2

Measure the effect of resources and environmental conditions on the size of a bacterial population.

1. Examine cultures of *E. coli* grown for 10 days in the following environments:
   - Distilled water, pH 7
   - Nutrient broth, pH 3
   - Nutrient broth, pH 5
   - Nutrient broth, pH 7
   - Nutrient broth, pH 9
   - Nutrient broth, pH 11
2. Quantify the relative turbidity of each culture between 0 (clear) and 10 (most turbid). Record your results in table 5.3.
3. If turbidometers are available, measure the turbidity of the solutions according to procedures demonstrated by your instructor. Record your results in table 5.3.

Table 5.3

Growth of Bacteria in a Limited-Nutrient Medium

Media	Turbidity Intensity (0–10)	Absorbance Value
Distilled water, pH 7	______	______
Nutrient broth, pH 3	______	______
Nutrient broth, pH 5	______	______
Nutrient broth, pH 7	______	______
Nutrient broth, pH 9	______	______
Nutrient broth, pH 11	______	______

Question 4

Compare your data for populations grown in nutrient broth and in distilled water. Does the presence of nutrients ensure rapid growth of bacteria? Why or why not? ____________

____

Procedure 5.3

Measure population growth of duckweed (*Lemna*).

1. During the first week of this term, your instructor placed 10 duckweed (*Lemna*) plants in an illuminated aquarium. Each week since then, he or she counted the number of plants in the aquarium. Those data are posted by the aquarium.
2. From now until the end of the term, count duckweed plants in the aquarium each week. Plot your data in figure 5.4.

Questions 5

What do you conclude about population growth of duckweed?______________________________

____

What will eventually happen to the size of the population? Why?______________________________

____

GROWTH OF HUMAN POPULATIONS

Our global population is growing extremely fast.

Procedure 5.4

Plot the historical growth of the human population.

1. Consider these data:

Year	Human Population (millions)	
8000 B.C.	5	
4000 B.C.	86	
A.D. 1	133	
1650	545	
1750	728	
1800	906	
1850	1130	
1900	1610	
1950	2400	
1960	2998	
1970	3659	
1980	4551	
1990	5300	
2000	6200	
2040	13,000	(projected)

2. Plot these data in figure 5.5.

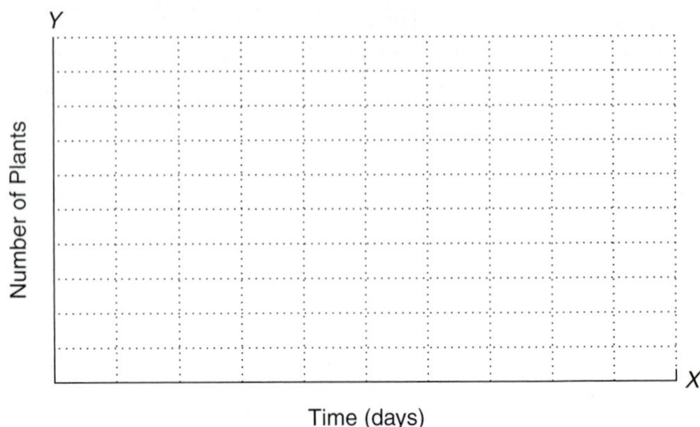

Figure 5.4
Population growth of duckweed (*Lemna*).

Questions 6

How does the shape of the graph in figure 5.5 compare with those you made for the bacteria? ____________

____

What do you conclude from your graph of human population growth? ____________

____

The population data listed in Procedure 5.4 have tremendous implications. For example, if our population had stabilized after World War II, today we could provide all of our energy needs (and have a higher standard of living) without having to burn any coal or import any oil.

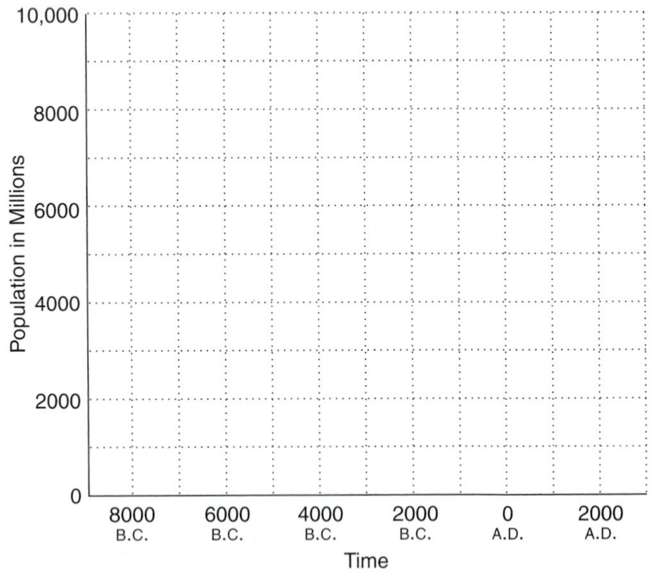

Figure 5.5
Growth of the human population.

Another important feature of a population is its **doubling time**. In 1850, the doubling time for the human population was 135 years. Today, the doubling time is about 40 years. Consequently, during that same 40 years we must also double our resources if we want to maintain our current standard of living. *Improving* our standard of living will require that we *more* than double our resources.

Questions 7

How does rapid population growth impact you? ____________

____

The doubling time for populations in developed countries is about 120 years but in developing countries it is about 30 years. What is the significance of this? ____________

____

Interestingly, the birthrate among Americans has climbed to its highest level since 1971 according to recent data (2006) from the National Center for Health Statistics. The birthrate hit 2.1 in 2006, which means that each female theoretically has 2.1 offspring. At this rate, each generation equally replaces itself. For industrialized countries, this is a rather high birthrate (table 5.4).

Table 5.4

Average Number of Births for Every Woman in Selected Developed Countries

Country	Birthrate
United States	2.1
France	2.0
Australia	1.8
United Kingdom	1.9
Germany	1.3
Russia	1.3
Japan	1.3

In contrast to growth in the United States, the global population has increased explosively during the past three centuries (fig. 5.6). Although the birthrate has remained constant (at about 30 per 1000), the death rate has fallen from about 30 per 1000 per year to its current level of about 13 per 1000 per year. The difference between birthrates and death rates (17 per 1000) means that the human population is growing at a rate of about 1.7% per year. Here's what that means:

- Each hour, the world's population grows by 11,000. Each year, the world's population grows by about 90,000,000.
- Each year, there are about 90 million more people on Earth. That annual increase in our population equals the combined population of Great Britain, Ireland, Iceland, Belgium, The Netherlands, Sweden, Norway, and Finland.

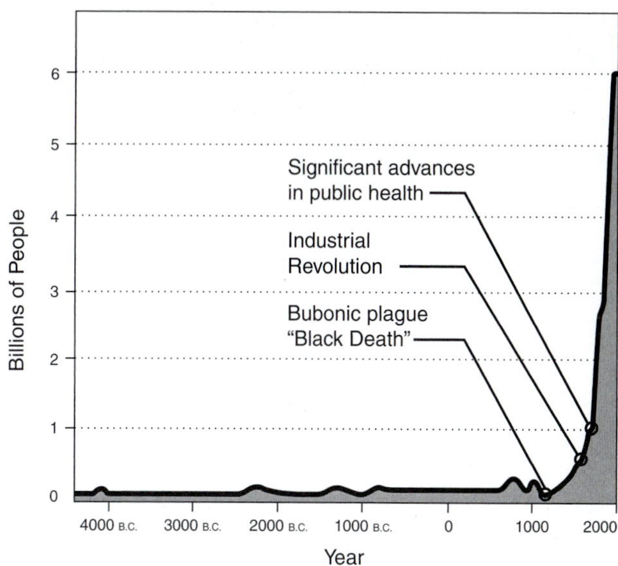

Figure 5.6

History of human population size. Temporary increases in death rate, even severe ones like the Black Death of the 1400s, have little lasting impact. Explosive growth began with the industrial revolution in the 1700s, which produced a significant long-term lowering of the death rate. The current population exceeds 6 billion, and at the current rate will double in 39 years.

- In the 6 seconds it takes you to read this sentence, 18 more people will be added to our population. Each of these people eats food, generates wastes, and, in his or her own way, affects our Earth.
- At our present rate of growth, in 2000 years our population will weigh as much as the entire earth. And 4000 years later, it would weigh as much as the visible universe.

Question 8

Can the current growth rate of humans continue? Why or why not? ____________

____

What will happen when our population exceeds the Earth's carrying capacity? ____________

____

How might the growth of the human population affect the growth of other populations? ____________

____

How does the growth of the human population affect ecosystems? ____________

____

Questions for Further Thought and Study

1. How can a population be slowed by its own numbers?

2. Some people are now realizing the significance of population growth. Although this exercise treated the problem only in biological terms, the reality of population growth is much more complex because it involves political, social, and economic problems. What are some of these problems? How do they affect you now? How will they affect you later in life (e.g., when you want to retire)?

3. Should we do anything to slow the "population explosion"? If so, what? If not, why?

4. From a purely ecological standpoint, can the problem of world hunger ever be overcome by improved agriculture alone? What other components must a hunger-control policy include?

5. How are problems such as deforestation, pollution, and world hunger linked to population growth?

6. The late Garrett Hardin, a famous biologist, wrote that "Freedom to breed will bring ruin to us all." Do you agree with him? Explain your answer.

Age Distribution and Survivorship

6

Objectives

As you complete this lab exercise you will:

1. Construct and examine age pyramids for growing and nongrowing populations.
2. Construct survivorship curves for organisms with contrasting life histories.
3. Gather and use cemetery demography data to assess the survivorship, mortality, and age distribution of a local human population.

Most natural populations include members of different ages. Maturing individuals encounter mortality factors and survivorship rates that vary among young, old, and all ages in between. To examine survival from birth to death, ecologists assemble age distribution data in a fairly standardized format called a **life table**. Life tables are constructed with rows of data for each age interval from birth to death. The initial data typically include the percent of the population in each age class and allow us to calculate survival rates (and reciprocal mortalities). The ultimate value of life tables is to reveal which life stages are experiencing mortality pressures and to assess the potential for population growth.

Questions 1

Are populations that suffer early and high rates of mortality necessarily limited in their growth? ________________

____

How might a population adapt or compensate for high mortality of its young stages? ________________

____

During which life stages of a large mammal such as a bear would you suspect that mortality factors such as disease and predation are highest, and therefore survival is lowest?

____

____

Plants offer no parental care beyond a seed's food supply and seed coat protection. What effect would a lack of extended parental care have on early survivorship? ________________

____

In this lab exercise you will examine and graph age distributions. Then you'll gather data sets of birth and death information from a cemetery to construct survivorship curves of a past human population.

AGE DISTRIBUTION

The number of individuals in each age class quantifies a **population age distribution**. Age distributions interest ecologists because they reveal much about a population's potential for change and which stages of life are most subject to mortality factors.

A population's age distribution can be graphed as the percent of the population occupying each age class (figure 6.1). This graph is sometimes called an **age pyramid** because it typically has a broad base of young individuals and a narrow apex of old individuals.

Questions 2

For each of the age pyramids in figure 6.1, what age classes include the youngest 50% of the entire population? ______

____

____

The death rate of the population of Sweden equals the birthrate, and growth is zero. What characteristics of its age pyramid reflect that fact? ________________

____

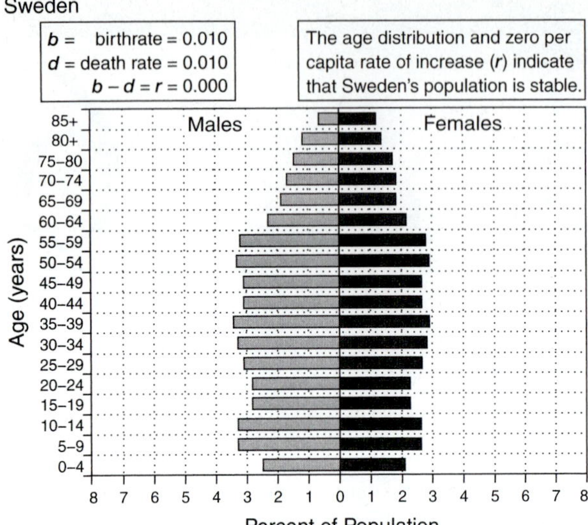

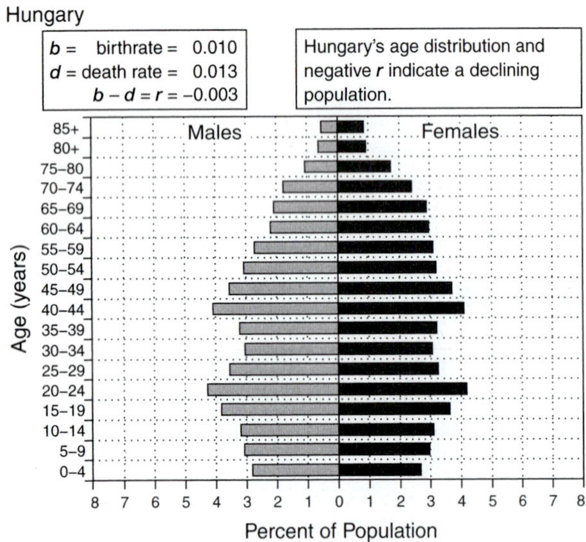

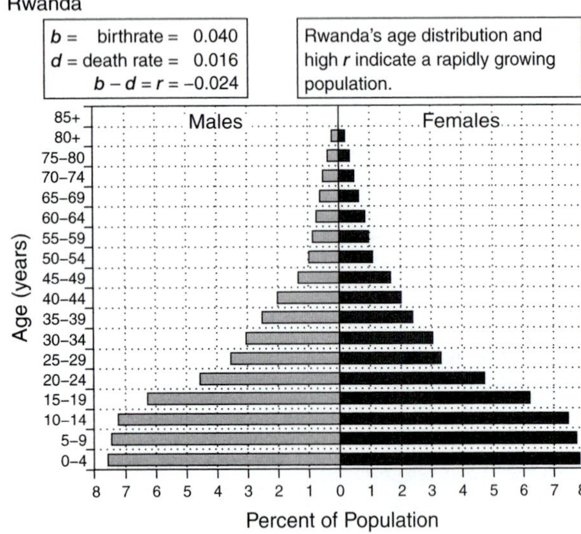

Figure 6.1

Age distributions for human populations in countries with stable, declining, and rapidly growing populations. Data from the U.S. Bureau of the Census, International Database 2006.

6–2

Procedure 6.1

Construct age pyramids for the age distribution data of three known populations.

1. Examine the three data sets presented in table 6.1. Assume that the data provided is only for females, and that males are equal in number.
2. For each data set calculate and record in table 6.1 the *Percent in Each Age Class*.
3. Construct in figure 6.2 an age pyramid for each of the data sets in table 6.1. Plot *Age Class* vs *Percent in Each Age Class*.

Questions 3

Which of the three pyramids shown in figure 6.2 has the broadest base? ____________________

____

How do these pyramids compare to those for human populations shown in figure 6.1? ____________________

____

AGE-SPECIFIC SURVIVORSHIP

With each passing age interval, individuals encounter mortality factors associated with that age. Not all individuals alive at the beginning of an age class will survive to the next class. In other words, survival rate is age specific. Ecologists portray age-specific survivorship with a **survivorship curve** (figure 6.3). For this graph, the *log_{10} of the number of survivors* is the dependent variable on the y axis. *Age* divided into age classes is the independent variable on the x axis.

Survivorship curves provide an informative view of a lifetime of varying survivorship rates. The slope of the curve at any age class reflects survivorship. For some species, survival is high during the early and mid stages of life and low (high mortality) late in life. Other species experience roughly the same mortality and survivorship throughout life. Still other populations experience high mortality during early age classes. These life-history strategies produce type I, II, and III curves, respectively (figure 6.4).

Questions 4

Which type of survivorship curve is typical for a human population? ____________________

____

Would you expect the survivorship curves to vary between developed and undeveloped countries? How so? __________

____

Survivorship rates are calculated from age distribution data gathered either by (1) following a group, or **cohort**, of "new-born" individuals (age class 0) as they pass through successive age classes; or (2) counting all individuals in each age class in a single **static** observation. To help compare populations of different sizes, the raw counts for each age class are usually standardized as a proportion of 1000 individuals at the beginning of age class 0.

Table 6.1

THREE DATA SETS OF AGE DISTRIBUTIONS INCLUDING AMERICAN ROBINS (FARNER, 1945), DALL MT. SHEEP (DEEVEY, 1947), AND SIMULATED DATA FOR OAK TREES

Robins			Dall Mt. Sheep			Oak Simulated Data		
Age Class (yrs) x	Observed Individuals a_x	Percent in Each Age Class	Age Class (yrs) x	Observed Individuals a_x	Percent in Each Age Class	Age Class (yrs) x	Observed Individuals a_x	Percent in Each Age Class
0	303		0	1000		0	1246	
1	150		1	801		1	67	
2	69		2	789		2	12	
3	30		3	776		3	3	
4	11		4	764		4	3	
5	3		5	734		5	2	
6	2		6	688		6	1	
7	0		7	640		7	0	
			8	571				
			9	439				
			10	252				
			11	96				
			12	6				
			13	3				
			14	0				

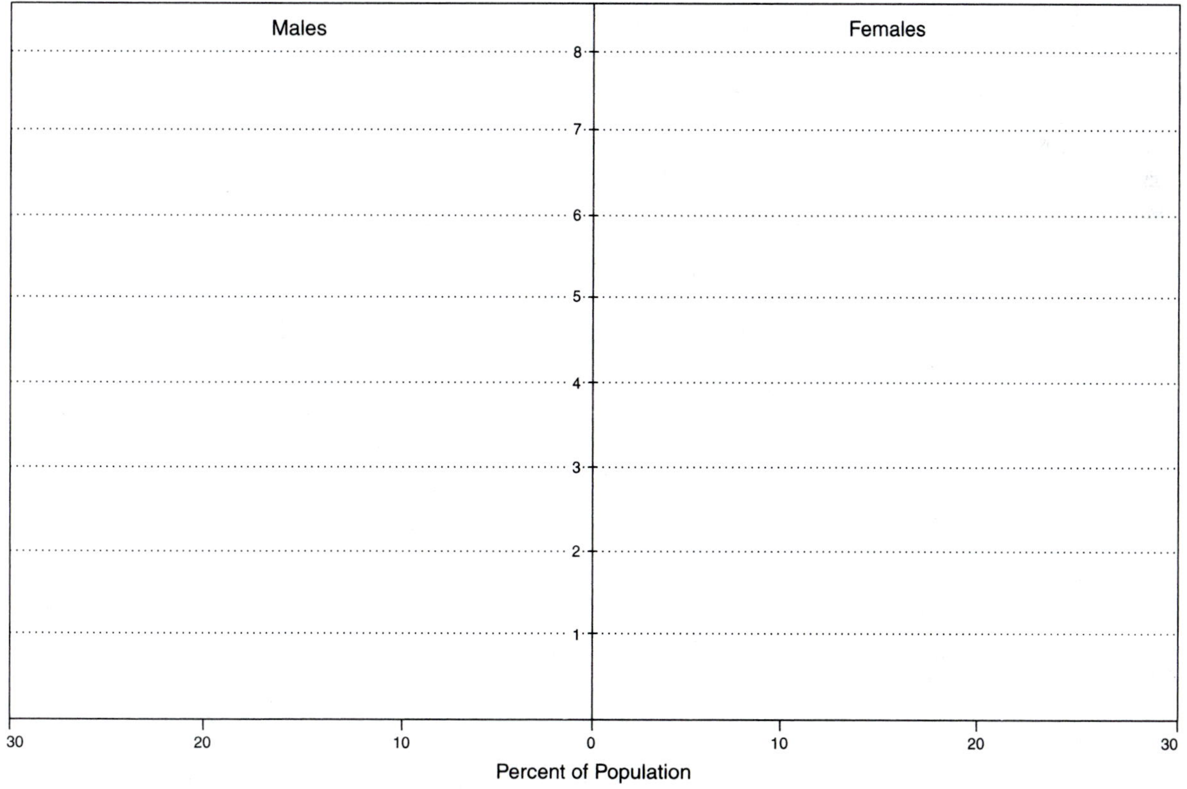

Figure 6.2a

Labeled axes for student-constructed age pyramid for robins (data from table 6.1).

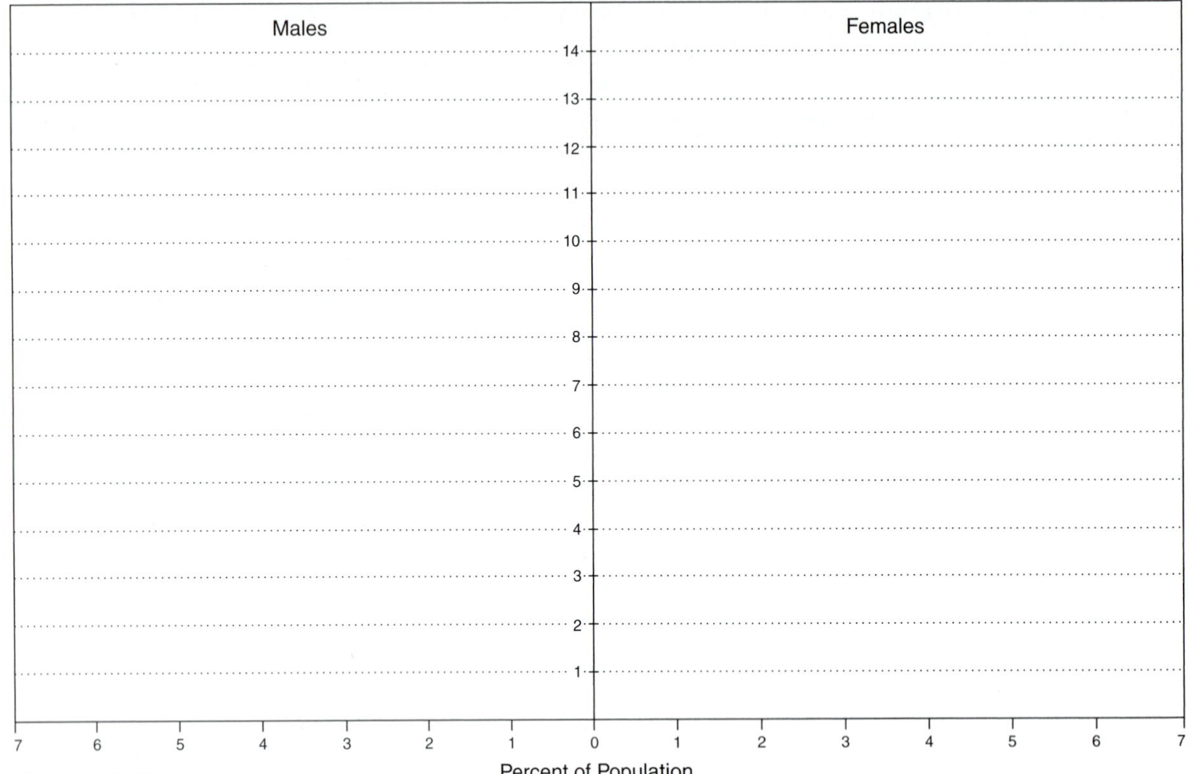

Figure 6.2b

Labeled axes for student-constructed age pyramid for mt. sheep (data from table 6.1).

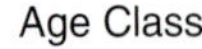

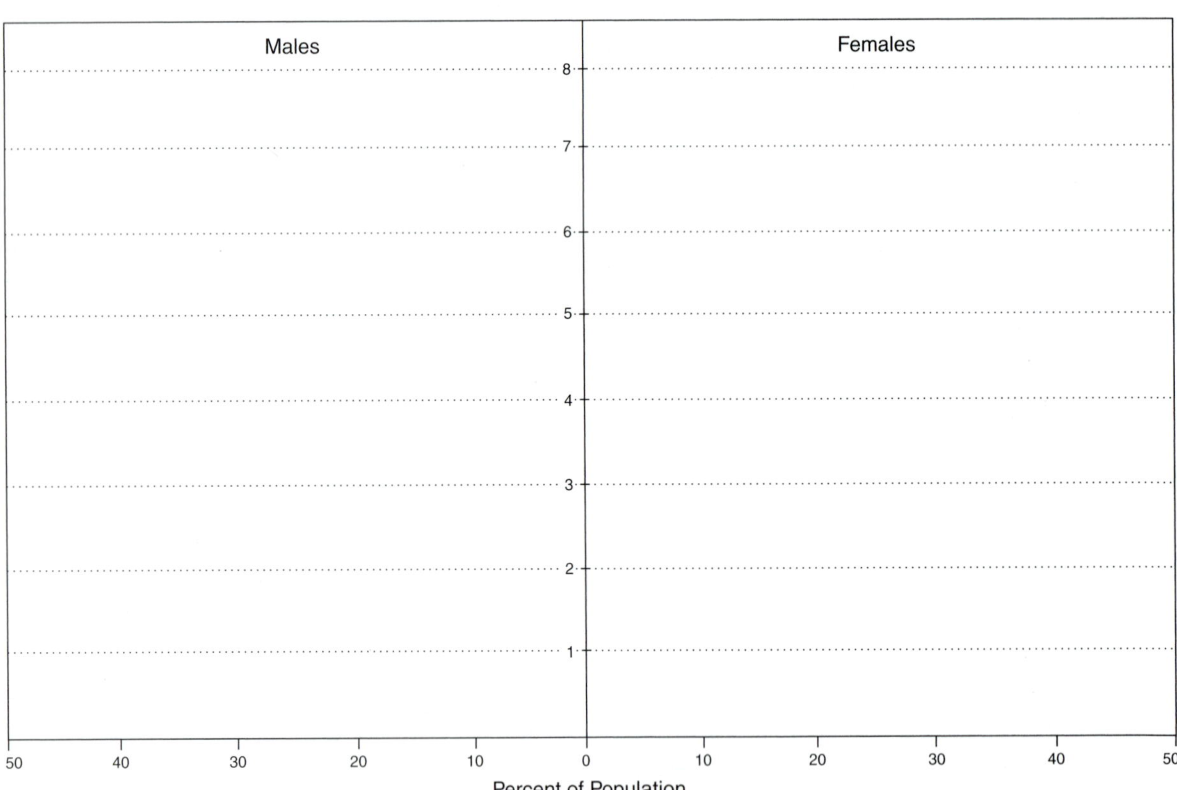

Figure 6.2c

Labeled axes for student-constructed age pyramid for oak seedlings (data from table 6.1).

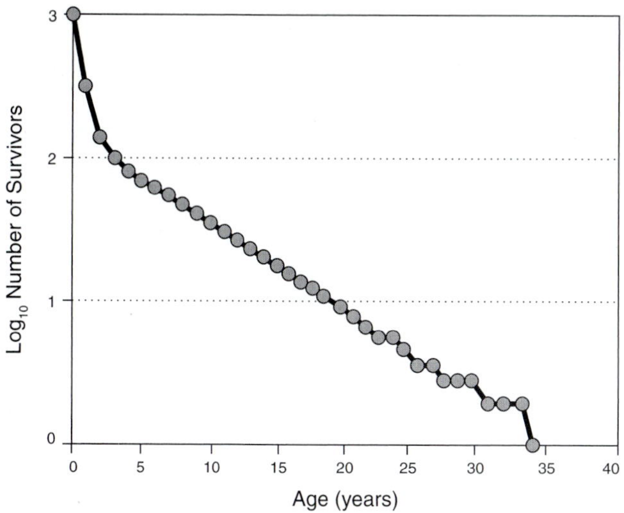

Figure 6.3
Constant rate of survival for a common mud turtle population (data from Deevey 1947, Baker, Mewaldt, and Stewart 1981, Frazer, Gibbons, and Greene 1991).

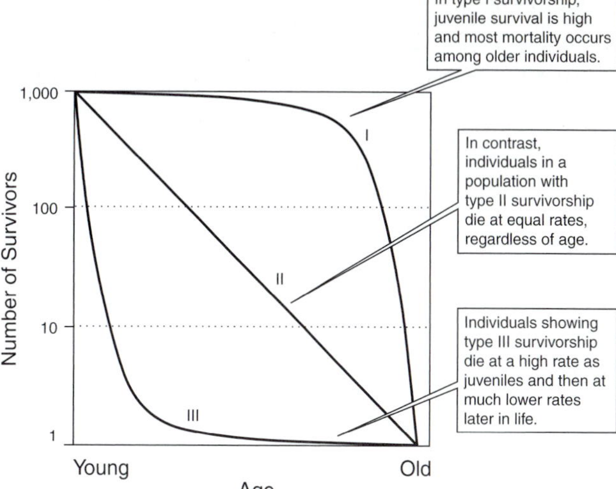

Figure 6.4
Theoretical types I, II, and III survivorship curves.

Question 5
Which of the two methods of gathering age and survivorship information assumes that the mortality rates remain stable through time? ____________________

____

Procedure 6.2

Plot survivorship curves for three populations.

1. Examine the age distribution data in table 6.2. The a_x column is the number of individuals alive at the beginning of the age class.
2. The n_x column is the number of individuals alive at the beginning of each age class. These numbers are standardized to 1000 for the first age class. Calculate and record each n_x value as:

   $n_x = (a_x / a_0) \cdot 1000$

   For example, n_x for age class 1 is $(150/303) \cdot 1000 = 495$.
3. Calculate and record in table 6.2 the $log_{10} n_x$ for each age class.
4. Construct in figure 6.5 a survivorship curve for each of the three data sets by plotting $log_{10} n_x$ versus *age class*.

Questions 6
Examine the slope of each segment of the three survivorship curves. During which year of the first 5 years is mortality the greatest for each species? ____________________

____

During which of the first 5 years is mortality the least?

____

____

Which of the three species has a type I curve? Type II curve? Type III curve? ____________________

____

CEMETERY DEMOGRAPHY

Survivorship curves reveal changing survival and mortality rates during a lifetime. Survivorship data for human populations are easily gathered because we leave behind accurate records of birth and death dates on gravestones in cemeteries. Dates on these gravestones allow us to track survivorship of cohorts of humans that passed through decades of mortality.

To gather data and construct survivorship curves for a local human population, your instructor has selected one or more cemeteries with gravestones and birth dates routinely dating back 100 years. You will work in teams and gather data for two cohorts. A cohort is a group of humans born in the same decade. Your class will gather birth year, death year, and gender information from ≈ 100 grave stones for the birth dates in the 1870s and 1890s.

TABLE 6.2

THREE AGE DISTRIBUTIONS INCLUDING DATA FOR AMERICAN ROBINS (FARNER, 1945), DALL MT. SHEEP (DEEVEY, 1947), AND A SIMULATED POPULATION OF OAK TREES

Robins				Dall Mt. Sheep				Oak Simulated Data			
Age Class (yrs) x	Observed Individuals a_x	Individuals Alive at Beginning of Age Class (standardized to 1000) n_x	$\log_{10} n_x$	Age Class (yrs) x	Observed Individuals a_x	Individuals Alive at Beginning of Age Class (standardized to 1000) n_x	$\log_{10} n_x$	Age Class (yrs) x	Observed Individuals a_x	Individuals Alive at Beginning of Age Class (standardized to 1000) n_x	$\log_{10} n_x$
0	303	1000	3.00	0	1000	1000		0	1246	1000	
1	150			1	801			1	67		
2	69			2	789			2	12		
3	30			3	776			3	3		
4	11			4	764			4	3		
5	3			5	734			5	2		
6	2			6	688			6	1		
7	0			7	640			7	0		
				8	571						
				9	439						
				10	252						
				11	96						
				12	6						
				13	3						
				14	0						

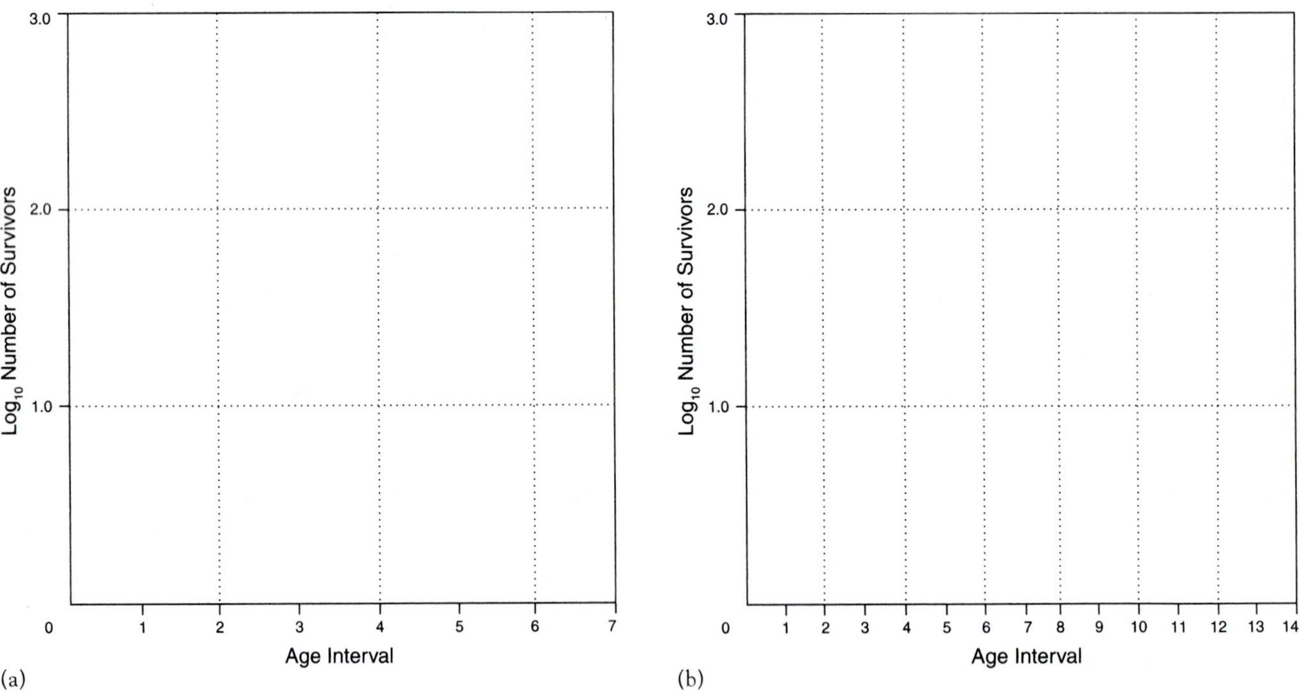

Figure 6.5

Student-constructed survivorship curves for (a) robins; (b) Dall mt. sheep; and (c) oak simulated data. Data are from table 6.2.

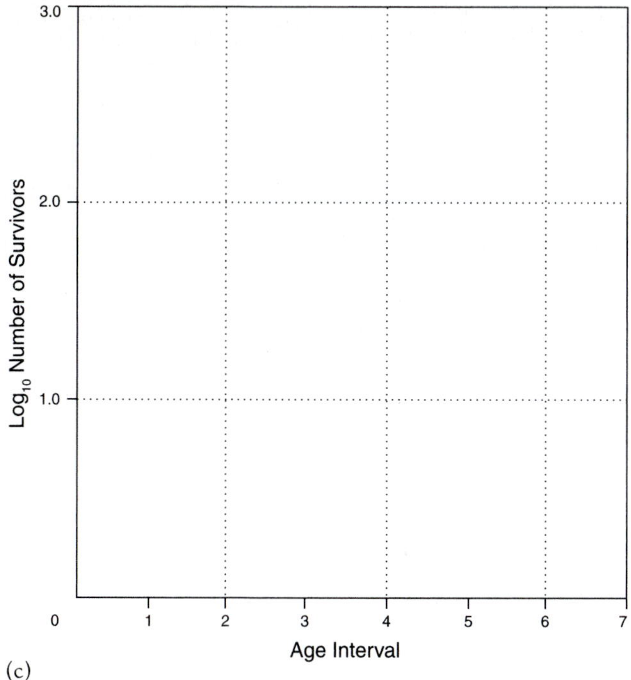

(c)

Procedure 6.3

Gather cemetery demography data.

Survey gravestones

1. Locate a large cemetery with gravestones showing birth dates in the 1870s and 1890s. These decades include the cohorts you will follow through time. Examine table 6.3 for recording your raw data. Make multiple copies of table 6.3 for all members of your team.
2. Divide into eight teams. Your instructor will assign four teams to each of the two cohorts (decades) investigated. Each team will record data from 25 gravestones.
3. Walk the cemetery and read gravestones to find individuals born in the decade your team was assigned. Record in table 6.3 the *birth year, death year,* and *gender* of 25 people born in that decade. If the gender is not apparent by the first name, then skip that gravestone. Don't count gravestones that other teams have already recorded.
4. On your raw data sheet (table 6.3), calculate and record for each individual his or her *Age at Death* by subtracting *Birth Year* from *Death Year.*
5. Examine data summary tables 6.4, 6.5, and 6.6. Notice that each age class includes 10 years.
6. For table 6.4, calculate the number of deaths during each age class (i.e., d_x) and record them in the column for your cohort and your team number. For example, to determine d_x of *Age Class* 0–9, count the number of individuals who died between 0 and 9 years of age.
7. Repeat step 6, but record in table 6.5 only the data for males.
8. Repeat step 6, but record in table 6.6 only the data for females.

Summarize the data from class teams

9. For tables 6.4, 6.5, and 6.6, record values for the remaining three d_x columns in each of the tables from the summary data sheets of the other three teams.
10. For each *Age Class*, sum across the four values from the four teams and record the total for the *Age Class* in column *TOTAL Number of Deaths in Age Class* d_x. Repeat for tables 6.4, 6.5, and 6.6.
11. Sum down the column the values of the *TOTAL Number of Deaths in Age Class* d_x. For table 6.4, it should equal 100 (i.e., the total number of gravestones recorded by all four teams). Record this value for a_x of *Age Class* 0–9.
12. Repeat step 11 for tables 6.5 and 6.6. The sum, however, will not equal 100 for either table.
13. For tables 6.4, 6.5, and 6.6, calculate the remaining values for *Observed Number Alive at the Beginning of the Age Class* a_x values as:

$$a_x = a_{x-1} - d_{x-1}$$

14. For tables 6.4, 6.5, and 6.6, calculate values for the a_x Standardized to 1000 n_x column as:

$$n_x = (a_x / a_{x-1}) \cdot (n_{x-1})$$

15. For tables 6.4, 6.5, and 6.6, calculate values for the *SURVIVORSHIP CURVE* log_{10} (n_x) column.
16. Exchange data with the teams working on the other cohort.

Plot survivorship for each cohort and gender

17. Plot two survivorship curves (combined genders) in figure 6.6. One curve is for the 1870 cohort, and one is for the 1890 cohort.
18. Plot two survivorship curves in figure 6.7. One curve is for 1870 females and one curve is for 1870 males.
19. Plot two survivorship curves in figure 6.8. One curve is for 1890 females, and one curve is for 1890 males.

Table 6.3

Raw data for cemetery demography

Team _______ Cohort _______

Record Number	Birth Year	Death Year	Gender	Age at Death
1				
2				
3				
4				
5				
6				
7				
8				
9				
10				
11				
12				
13				
14				
15				
16				
17				
18				
19				
20				
21				
22				
23				
24				
25				

6–8

Table 6.4

Summary life table for gender-combined data from cemetery demography

_____Cohort

Age Class (yr)	TEAM 1 Number of Deaths in Age Class d_x	TEAM 2 Number of Deaths in Age Class d_x	TEAM 3 Number of Deaths in Age Class d_x	TEAM 4 Number of Deaths in Age Class d_x	TOTAL Number of Deaths in Age Class d_x	Observed Number Alive at the Beginning of the Age Class a_x	a_x standardized to 1000 n_x	SURVIVORSHIP CURVE $\log_{10}(n_x)$
0–9							1000	3.0
10–19								
20–29								
30–39								
40–49								
50–59								
60–69								
70–79								
80–89								
90–99								
100–110								

Table 6.5

Summary life table for the male survivorship data

_____Cohort

Age Class (yr)	TEAM 1 Number of Male Deaths in Age Class d_x	TEAM 2 Number of Male Deaths in Age Class d_x	TEAM 3 Number of Male Deaths in Age Class d_x	TEAM 4 Number of Male Deaths in Age Class d_x	TOTAL Number of Male Deaths in Age Class d_x	Observed Number Males Alive at the Beginning of the Age Class a_x	a_x standardized to 1000 n_x	SURVIVORSHIP CURVE $\log_{10}(n_x)$
0–9							1000	3.0
10–19								
20–29								
30–39								
40–49								
50–59								
60–69								
70–79								
80–89								
90–99								
100–110								

Table 6.6

Summary life table for the female survivorship data

_____Cohort

Age Class (yr)	TEAM 1 Number of Female Deaths in Age Class d_x	TEAM 2 Number of Female Deaths in Age Class d_x	TEAM 3 Number of Female Deaths in Age Class d_x	TEAM 4 Number of Female Deaths in Age Class d_x	TOTAL Number of Female Deaths in Age Class d_x	Observed Number Females Alive at the Beginning of the Age Class a_x	a_x standardized to 1000 n_x	SURVIVORSHIP CURVE $\log_{10}(n_x)$
0–9							1000	
10–19								
20–29								
30–39								
40–49								
50–59								
60–69								
70–79								
80–89								
90–99								
100–110								

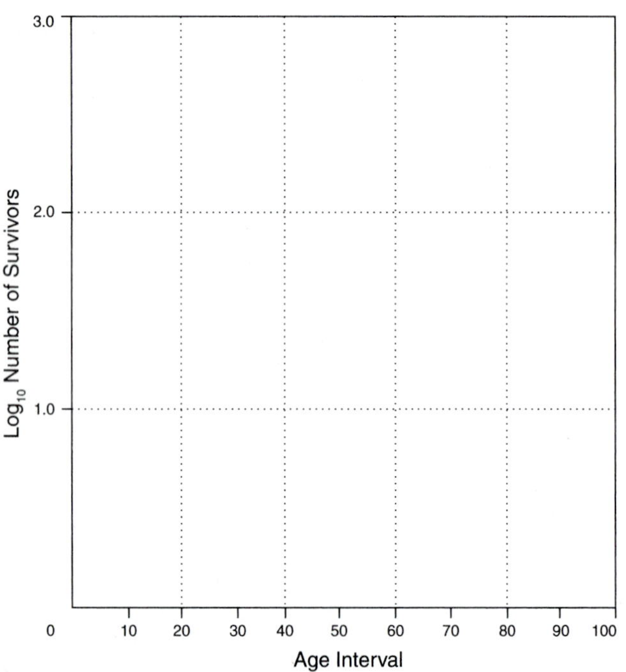

Figure 6.6

Survivorship curves for combined-gender. Data from table 6.4.

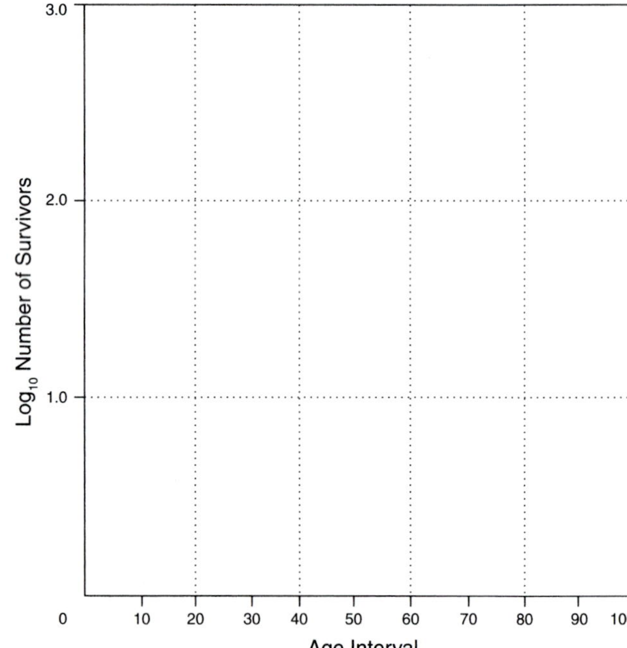

Figure 6.7

Survivorship curves for data for males from table 6.5.

6–10

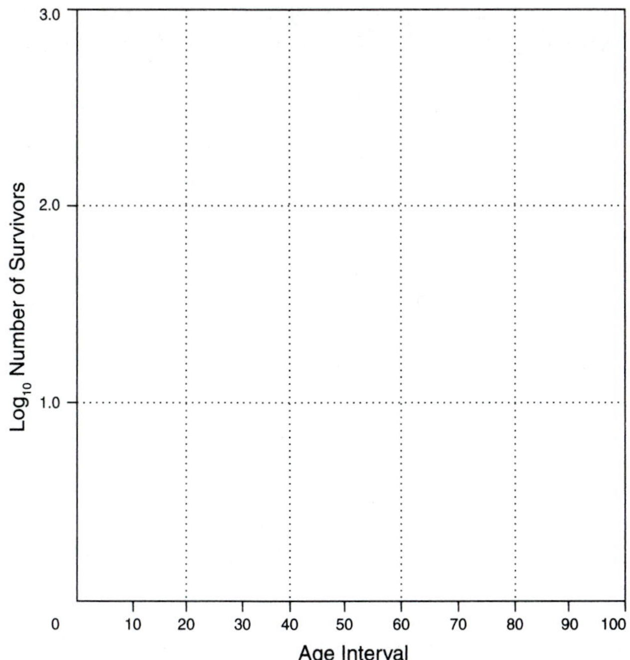

Figure 6.8

Survivorship curves for data for females from table 6.6.

Questions 7

Could a survivorship curve ever go up from one age interval to the next? Why? Or why not? ____________________

____

For which cohort did 50% of the population live the longest? ($\log_{10} 500 = 2.7$)____________________

____

How does the survivorship of males compare with that of females?____________________

____

Did the later cohort have longer survivorship? __________

____

What are some possible factors responsible for the different survivorship of the later cohort? ____________________

____

During which age intervals was survivorship the greatest?

____

____

What are the weaknesses inherent in using a single cemetery to characterize survivorship of a population? ________

____

Questions for Further Thought and Study

1. For what kinds of species would longevity correlate with population growth? For what kinds would it not correlate?

2. During which age interval would life-saving medical advances have the most impact on potential population growth? Why?

Terrestrial Plant Community Assessment 7

Objectives

As you complete this lab exercise you will:

1. Characterize a terrestrial community as to observable physical factors, plant dominance, and interactions.
2. Quantify the distribution and abundance of plants in a community.
3. Use transect data to quantify the importance of different plant species in a plant community.

Ecological communities are extraordinarily complex. The plant community you observe at any given time is the product of interactions among (1) plants and their physical surroundings; (2) different species of plants; and (3) plants and animals (fig. 7.1). The engine driving these interactions is the flow of energy captured by green plants and passed to consumers and decomposers.

This exercise cannot explain all of the processes occurring in a plant community, but it can guide you through some basic observations that characterize and distinguish communities. Do not underestimate the importance of qualitative observations in understanding communities followed by designing quantitative studies to address specific hypotheses.

QUALITATIVE COMMUNITY ASSESSMENT

Procedure 7.1

Observe and assess the ecological characteristics of a terrestrial community.

1. Locate and visit a terrestrial community designated by your instructor.
2. Characterize the community according to the criteria and questions that follow. After you've answered the questions, discuss your observations with your instructor and other groups. Be prepared to use your observations as a basis for describing (1) your assessment of energy flow through the community; (2) the diversity of the community; and (3) the interactions among organisms.

Figure 7.1
This terrestrial community is a diverse and interacting mix of plants, animals, and microorganisms. They interact with the abiotic environment and compete for nutrients, light, moisture, and shelter.

PHYSICAL FACTORS

Observations

1. What levels of light intensity occur throughout the community? ______________________

____

2. Does the community include shade-tolerant as well as shade-intolerant plants (i.e., are some plants doing well in the shade and some better in full sunlight)?

____

____

3. How does light differ among vertical levels of vegetation? ______________________

____

4. What is the temperature 2 m above ground? _______

5. What is the temperature at the soil surface? _______

6. How much and in what direction does the ground slope? _______

7. How would you characterize the soil? Loam? Clay? Sand? _______

8. What is the nature of the groundcover? Grasses? Bare soil? _______

9. Is there a layer of leaf litter on the ground? _______

10. Is the environment generally moist, moderate, or dry? _______

Interpretations

1. How might shade affect the temperature of the community? _______

2. How might different amounts of light at different vertical levels within the community be important? _______

3. What parts of the community might be cooler than others? Why could this be important? _______

4. Why would ground slope be important? _______

5. Based on your observations of slope and soil type, would you expect the soil to retain moisture? _______

6. How long has this community been left to develop without disturbance? That is, what is the apparent age of the community? _______

PLANT DOMINANCE

Observations

1. Most plant communities are dominated by one, two, or three species. Is the plant community dominated by a single species? Two species? _______

2. Which plant species are most abundant (numbers)? _______

3. Which plant species are most abundant (biomass)? _______

4. What general categories of plant types (shrubs, trees, etc.) are apparent? _______

5. What is the vertical distribution of vegetation? _______

Interpretations

1. Would you describe this community as diverse? Why or why not? _______

2. What comparable community in your local area would you consider to be more diverse? Less diverse? _______

3. What observations led to your conclusion for the previous question? _______

4. Are there specific factors that make your comparison community more or less diverse? Human impact? Stressful environmental factors? Geology? _______

INTERACTIONS AMONG ORGANISMS

Observations

1. What evidence do you see of resident vertebrates? _______

2. What evidence do you see of resident invertebrates? _______

3. What evidence do you see of plant-animal interactions? _______

4. What evidence do you see of plant-plant interactions? _______

7–2

5. What adaptations do the plants have to discourage herbivores? ______________________________

____

6. Do you see any obvious or subtle evidence of competition by plants for available resources? ______

____

Interpretations

1. If you don't see any vertebrates, does that mean they are not around? Explain your answer. ____________

____

2. Reexamine the observations that you just listed. How would each observation affect the type and growth of plants in the community that you studied? ________

____

3. What kinds of competitive interactions are apparent in the community? ______________________

____

4. What kinds of mutually beneficial interactions are apparent in the community? ________________

____

QUANTITATIVE COMMUNITY ASSESSMENT

Ecologists have developed a variety of techniques to measure the numbers, densities, and distributions of organisms in terrestrial plant communities. One common technique is to count organisms within randomly distributed **quadrats** (sometimes called **plots**) of uniform size. See Exercise 10 for more information about using quadrats. Another common technique is the **line-intercept method**. In this method, a **transect**, or line, is laid out within the community. Organisms in contact with this line are counted and measured. Calculations based on measurements from these line transects or quadrats reveal the relative abundances, frequencies, and distributions of the plant species that compose the community.

Procedure 7.2

Assessing a community with the line-intercept method.

1. With the help of your instructor, locate a suitable field site with a plant community to be examined.
2. Obtain a measuring tape 10–15 m long, a meter stick, and a notepad. If a measuring tape is unavailable, use a measured piece of string or rope.
3. Assess the general layout of the community to be sampled. With the aid of your instructor, decide on a reasonable set of criteria to govern the placement of a transect for each group of students. One common method is to establish a baseline along one side or within the community. Randomly choose points along this line as starting points to lay out perpendicular transects.

Questions 1

What concepts or ideas should govern the placement of your transect to obtain a representative sample of the community? ______________________________

____

Are there any "wrong" places to put a transect? Why or why not? ______________________________

____

4. You and your lab partners will work on a single transect. Stretch the measuring tape on the ground to establish a transect.
5. Divide the transect into 1-m or 5-m intervals to facilitate frequency calculations.
6. At the top of table 7.1 record the total transect length (L_{total}) and total number of intervals (I_{total}).
7. For the first interval, identify plants that touch, overlie, or underlie the transect line. Treat bare ground as a "species."
8. In your field notes record each type (species) of plant. Also, for each species record the total length of the line intercepted by all individuals of that species. For plants that overhang the line, record the length of the line's imaginary vertical plane that the plant intercepts.
9. Repeat steps 7 and 8 for each transect interval.
10. When all plants from all intervals have been recorded in your field notes, summarize your data in table 7.1.
11. Sum the values in each of the three data columns of table 7.1 to calculate N_{total}, F_{total}, and C_{total}. Record the calculations at the bottom of each column.
12. Use the data in table 7.1 to calculate *Density* and *Relative Density* for each species within the community. Record your results in table 7.2.

$$Density_i = n_i / L_{total}$$
$$Relative\ Density_i = n_i / N_{total}$$

13. Use the data in table 7.1 to calculate the following *Frequency* and *Relative Frequency* for each species within the community. Record your results in table 7.2.

$$Frequency_i = f_i / I_{total}$$
$$Relative\ Frequency_i = f_i / F_{total}$$

14. Use the data in table 7.1 to calculate *Coverage* and *Relative Coverage* for each species within the community. Record your results in table 7.2.

$$Coverage_i = c_i / L_{total}$$
$$Relative\ Coverage_i = c_i / C$$

15. Use the data in table 7.1 to calculate the *Importance Value* for each species within the community. Record your results in table 7.2.

importance value of species$_i$ = *relative density*$_i$ + *relative coverage*$_i$ + *relative frequency*$_i$

Questions 2

What is the meaning of an importance value? __________

____

Why would we calculate this in addition to density, coverage, and frequency? __________

____

Table 7.1

Summary of data for plant species occurring along a transect

L_{total} = total length of transect = __________ I_{total} = total number of intervals = __________

Species$_i$	n_i = Total Number of Individuals Encountered for Entire Transect	f_i = Number of Intervals in Which Species *i* Occurs	c_i = Total Length of Transect Intercepted for All Intervals
	N_{total} = Total of all individuals = ____	F_{total} = Total of all frequencies = ____	C_{total} = Total length of transect intersected = ____

Table 7.2

Relative values of each species in a selected community using parameters of the line-intercept method

Species$_i$	Density	Relative Density	Frequency	Relative Frequency	Coverage	Relative Coverage	Importance Value

Questions for Further Thought and Study

1. Diverse plant communities have species representing a variety of plant types such as grasses, shrubs, succulents, hardwood trees, softwood trees, vines, ferns, and so on. What factors increase a community's diversity? Age of the community? Energy input? Moisture? Nutrients? Disturbance? Human activity? How do they do so?

2. What characteristics of a community might make it more resilient than other communities after disturbance?

3. What characteristics indicate that a community has not been disturbed for a few years?

Stream Ecosystem Assessment

8

Objectives

As you complete this lab exercise you will:

1. Define a stream's *drainage basin* and explain how it is influenced by surrounding land.
2. Prepare a linear morphometric map of the stream.
3. Measure and record water temperature, water velocity, and stream discharge along the stream channel and shoreline.
4. Assess sediment particle size, primary producers, benthic invertebrates, and fish populations for riffles versus pools.

Rivers and streams drain rain and melting snow from terrestrial ecosystems. This runoff water eventually collects in small rivulets that join to form a network of channels that drain the landscape. Quiet **pools** in the channels may have a **current velocity** of only a few millimeters per second, whereas water in the **riffles** may flow at 6 m per sec. (fig. 8.1). This continuous movement of flowing water is the most prominent characteristic of a stream. Its current delivers food, removes wastes, renews oxygen, and strongly affects the size, shape, and behavior of organisms. The volume of water per hour flowing past a point of a stream is its **discharge**.

A stream **basin** is the area of land drained by a network of rivers and streams. Streams and rivers can be classified by **stream order** according to where they occur in the drainage network. Headwater streams are first order. A stream formed by the joining of two first-order streams is second order. A third-order stream results from the joining of two second-order streams and so on. A lower-order stream joining a higher-order stream does not raise the order of the stream below the junction (fig. 8.2).

Streams are more strongly impacted by their surrounding environment than you think. First- and second-order streams are generally shaded by **riparian vegetation** occurring along the shorelines and forming a transition between the aquatic and terrestrial environments. Shading may be so thorough along some streams that little photosynthesis by aquatic primary producers occurs. Shading lessens downstream as stream width increases.

Figure 8.1
Pools and riffles of a stream are formed by variation in sediment erosion and deposition. Although they share the same flowing water, these microenvironments vary in current speed, substrate type, and resident species. This photo shows pools separated by rocky, shallow riffles.

In this lab exercise you will assess some major physical and biotic characteristics of a local stream segment. You'll characterize the pattern of surrounding land use, as well as flow dynamics, sediment variation, invertebrate communities, and fish communities within riffles and pools.

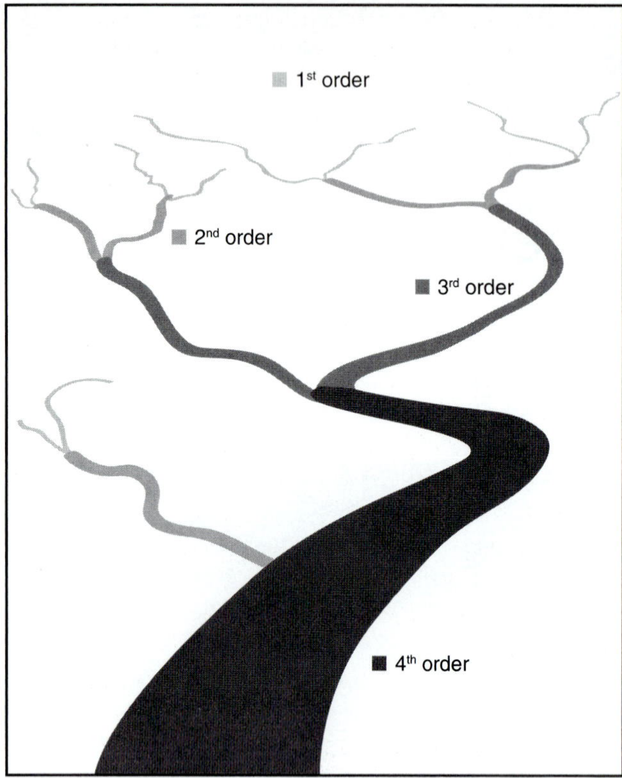

Figure 8.2

A drainage network illustrating stream-order classification for a fourth-order watershed.

PHYSICAL CHARACTERISTICS

Procedure 8.1

Define the drainage basin and the influence of surrounding land.

1. Locate on a county map a stream with an accessible 50-m sampling site, and identify the surrounding streams and rivers.
2. Follow on the map the upstream path of the stream from your sampling site, and generally define the boundaries of the stream's drainage basin.
3. Use the scale on your map and the locations of the surrounding rivers and streams to estimate the drainage basin area in square kilometers (or miles).
4. Determine the land use in the drainage basin either by discussing the area with your instructor, locating a map that shows land-use patterns, or driving through the area.
5. Determine the land-use pattern within the 150-m wide border along each side of your sampling area.
6. Survey the riparian vegetation. The riparian vegetation (vegetation directly influenced by the stream) may extend as far as 20 m from the water's edge.
7. Identify three, four, or five of the most common riparian tree species, and count individuals of these species occurring per linear 10 m of stream reach. Count only trees with a trunk diameter of > 6 cm at breast height (DBH). **DBH** refers to the stem (or trunk) diameter at 1.5 m above the ground. For this exercise, the entire population of large trees along the stream reach can be counted. In Exercise 10 you will learn how to sample larger populations.

   species A, number per 10 m _______
   species B, number per 10 m _______
   species C, number per 10 m _______

8. Many streams are shaded by expansive and overgrown riparian vegetation. Walk the length of the sample area and estimate the areas of water's total surface receiving full sunlight, partial sunlight, and no direct sunlight.

   % area full sunlight _______
   % area partial sunlight _______
   % area no direct sunlight _______

9. If time permits, visit the stream 2 hours after sunrise, midday, and 2 hours before sunset to determine variation in how much of the water's surface receives full sunlight.

Questions 1

What order stream includes your sampling site? _________

____

What is the primary land use in the drainage basin? ______

____

Does the land use in the immediate area differ from that of the entire basin's? How so? _______________________

____

Which would have the greater impact on the stream's water quality: the drainage basin or the land immediately surrounding the stream sampling area? How so? ___________

____

Riparian vegetation is a buffer zone between the stream and the surrounding land. How extensive is the riparian buffer at your sampling site? ___________________________

____

What signs of erosion are apparent along the riparian zone?

____

____

Are the riparian trees different species from the trees away from the water's edge? ___________________________

____

Do you detect differences in vegetation among areas with different amounts of direct sunlight? How so? ___________

____

Rivers and streams often divide along their lengths into pools and riffles. Riffles have rapid flow and are shallow enough for the bottom sediment to cause noticeable turbulence. Pools are deeper areas of slower flow.

Procedure 8.2

Prepare a linear morphometric map of the stream.

1. Obtain graph paper with uniform boxes. Each box represents a scaled distance of 1 m or other value as determined by the instructor (fig. 8.3).
2. Draw a straight line spanning 50 units along the middle of the long axis of the paper. This line represents the middle of the stream at all points along the stream.
3. Begin at the upstream end of the 50-m stream reach and wade down the middle of the stream channel (equidistant from each shoreline). Face perpendicular to the channel and toward one shoreline. Direct a team member where to put a temporary flag marking the first of a series of 5-m intervals.
4. Wade 5 m farther down the center of the stream. Face the same shoreline and direct a team member where to put a temporary flag marking that 5-m interval.
5. Continue to walk the center line of the entire stream and mark 5-m intervals along the same shoreline. Bends in the stream result in markers closer together than 5 m on the inside of a curve, and more than 5 m apart on the outside of the curve. That's okay. They indicate a 5-m interval along the *center* of the stream.
6. Return to the upstream end of the stream and the first marker. Measure the width of the stream at that marker. Indicate on your map half that width (adjusted for scale) on either side of the center line. For example, a 10-m width has two points five boxes on either side of the end of the center line.
7. At the next shoreline marker, measure the width of the stream, and record that span across the center line of your map.
8. Repeat step 7 for all shoreline markers to the downstream end of the stream.
9. On your map, connect the dots along each side of the centerline to outline the two shorelines with a smooth line.
10. With your instructor's aid, identify riffles and pools. Shade and label the areas of your map that are obvious riffles or obvious pools.

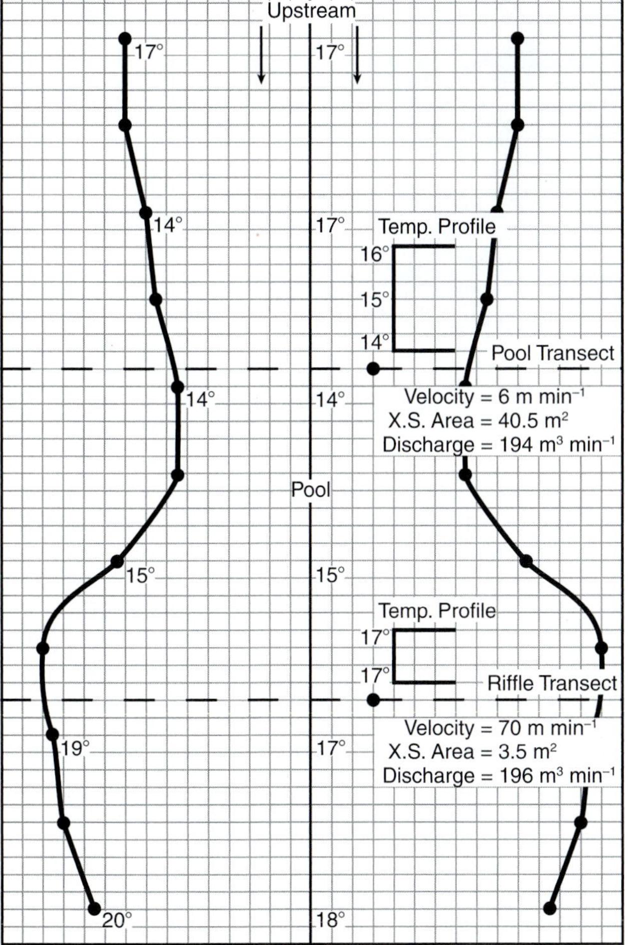

Figure 8.3

Example map of a stream segment. Temperatures at shoreline and midstream intervals are indicated, along with temperatures of two verticle profiles. The positions and characteristics of a pool transect and a riffle transect are indicated. Each block is 1 square meter.

Question 2

Do the riffles and pools locations relate to stream widths? How so? ______________________________

____

Procedure 8.3

Measure and record water temperature along the stream channel and shoreline.

1. Obtain an electronic thermometer with a probe (thermistor) that can be submerged (fig. 8.4). Measure and record on your map the air temperature.
2. In the middle of the channel across from the initial upstream marker, measure the temperature within 3–4 cm of the bottom. Record this value on your map at the upstream end of the centerline.
3. At the initial upstream marker, measure the water temperature at each shoreline. Place the thermistor within 0.3 m of the water's edges. Record these two values on your map.
4. Repeat steps 2 and 3 to take and record three temperature readings at each 5-m interval transect along the stream.

Figure 8.4
This digital thermometer has a thermistor probe at the end of a cable to lower to a known depth. This student is recording the temperature at the surface of shallow water.

5. Construct a vertical temperature profile of a pool. To do this, find the deepest part of the stream reach and take temperature readings at four equally distributed depths from surface to bottom. Indicate on your map the location measured, and record the values.
6. Construct a vertical profile of temperature for a riffle. To do this, record the temperature at the water surface and at the sediment surface. If possible, push the thermistor probe below the surface of the sediment and measure and record the temperature. Record on your map the location and temperatures.

Questions 3

Is there a downstream temperature gradient? __________

____

Why is shoreline water typically warmer than midchannel water? ______________________________________

____

Examine the shoreline temperatures closely. Do you detect any areas that the temperature is cooler than the midchannel water? What might account for this? ______________

____

Procedure 8.4

Measure and graph a depth profile for pools and riffles.

1. Obtain a meter stick, measuring tape, and graph paper divided into uniform blocks.
2. Review your stream map and locate the major riffles and pools.
3. With your instructor's guidance, choose a representative pool and a representative riffle appropriate for constructing a depth profile.
4. Measure the stream width from shoreline to shoreline across the middle of the pool. Determine an appropriate scale value for one block on your

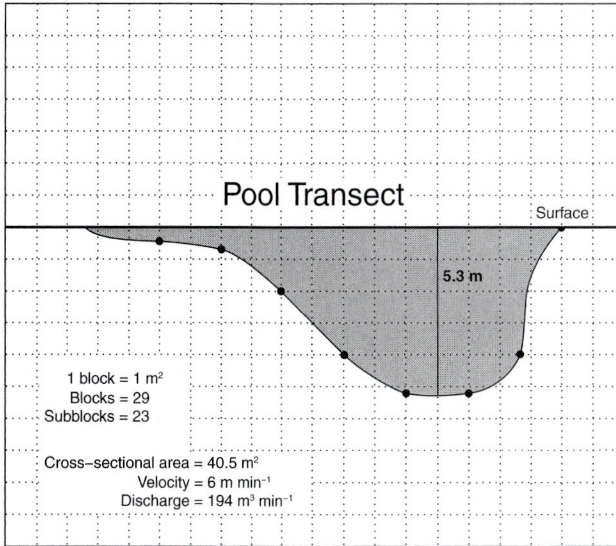

Figure 8.5

Example map of pool transect. Each block is 1 square meter. Depth readings were taken at 2-m intervals.

graph paper so the width of the pool spans the entire width of the graph paper (fig. 8.5).

5. Represent the stream's surface width as a horizontal line across the graph paper spanning the number of boxes appropriate to the map's scale.
6. Divide the width of the stream into 10–15 uniform intervals, and use the meter stick to measure the depth of water at each interval. Record each depth as a point the appropriate distance below the surface line on your depth-profile graph.
7. Map the bottom of the stream profile by connecting the points below the surface with a smoothly drawn line.
8. Calculate a transect's cross-sectional area of the stream:
   a. Determine the area represented by a single block. For example, if the side of a block represents 0.5 m, then one block is $0.5 \times 0.5 = 0.25$ m^2.
   b. Count the number of blocks completely enclosed within the lines of the stream cross section.
   c. Count the number of blocks that the bottom line subdivides.
   d. Sum the number of whole blocks (step b) and one half the number of intersected blocks (step c).
   e. Multiply the sum of blocks by the area of a single block (step a). Record this value on your depth-profile map as the cross-sectional area of the stream transect being measured.
9. Repeat steps 4–8 for the pool and riffle designated by your instructor.

Question 4

If the water entering the stream segment equals that amount of water leaving the segment, then would you expect the cross-sectional area to be the same at all points (transects) along the stream? Why or why not? ____________________

____

Procedure 8.5

Measure water velocity and stream discharge.

1. Locate the stream transects that were profiled in Procedure 8.4.
2. Obtain an orange (fruit) and a stopwatch. Verify that the orange floats with only a small portion above water. If not, try another orange.
3. At the first stream transect, mark a 5–10-m length of midchannel stream flow; the longer the segment, the better, as long as the stream dynamics are the same throughout the length.
4. Release the orange in the water a few meters upstream from the beginning of the segment. Start timing when it reaches the beginning of the segment. Stop timing when the orange reaches the end of the measured segment.
5. Calculate the velocity of water flow in meters per minute. Record this value on your linear stream map as well as on the depth profile for that transect.
6. Repeat steps 3–5, but measure velocity as close to the shorelines as possible rather than midchannel.
7. Calculate discharge (m^3 min^{-1}) by multiplying the cross-sectional area (m^2) of the stream transect (Procedure 8.4, step 8) times the midchannel velocity (m min^{-1}) (Procedure 8.5, step 5). Multiply this value by 0.8 to adjust for using only the midchannel velocity.
8. Record the discharge value alongside midchannel velocity measurements on your maps and depth profiles.
9. Repeat steps 3–8 for each transect profile that you constructed in Procedure 8.4.

Questions 5

Are midchannel velocities greater than velocities near the shoreline? Why or why not? ____________________

____

A rain event during the rainy season results in greater velocity and discharge than a rain event during the dry season. Why? ____________________

____

Are the discharge volumes at each of the transect profiles the same? Should they be? How so? ____________________

____

What might account for variation in discharge volumes from one part of the stream to another? ______________

____

Some streams are fed primarily by runoff, whereas others are fed by groundwater. How would this affect turbidity of the water? ______________________________________

____

Do you detect any areas of the stream that have significantly clearer water indicative of a groundwater seepage feeding the stream? ______________________________

____

The sediments of rivers and streams are more dynamic than you might expect. Inorganic and organic materials from the surrounding landscape continuously wash, fall, or blow into rivers. At the same time, flow and turbulence erode bottom sediments and keep them in suspension, particularly during floods. Areas of rapid flow, such as riffles, transport small sediment particles downstream and leave only the coarsest sediment. In pools, the slow current allows small particles to settle out of the water column.

Procedure 8.6

Assess sediment particle size for riffles versus pools.

1. Locate the stream transects with profiles and calculated discharges. Select a riffle transect and a pool transect.
2. If the typical particle size at the center of the riffle transect exceeds 1 cm, then randomly select 20–30 particles from a 0.25-m^2 plot (0.5 m × 0.5 m), measure their greatest dimension, and determine the median size.
3. If the typical particle size at the center of the riffle transect is less than 1 cm, then use a trowel to remove the sediment to a depth of 6 cm from a 400-cm^2 plot (20 cm × 20 cm) in the middle of the stream channel. Place the sediment in a bucket.
4. Rinse the collected sediment through a series of stacked sediment sieves to separate it into size classes (fig. 8.6).
5. Examine the sieves and record in table 8.1 the sequence of mesh sizes from largest to smallest.

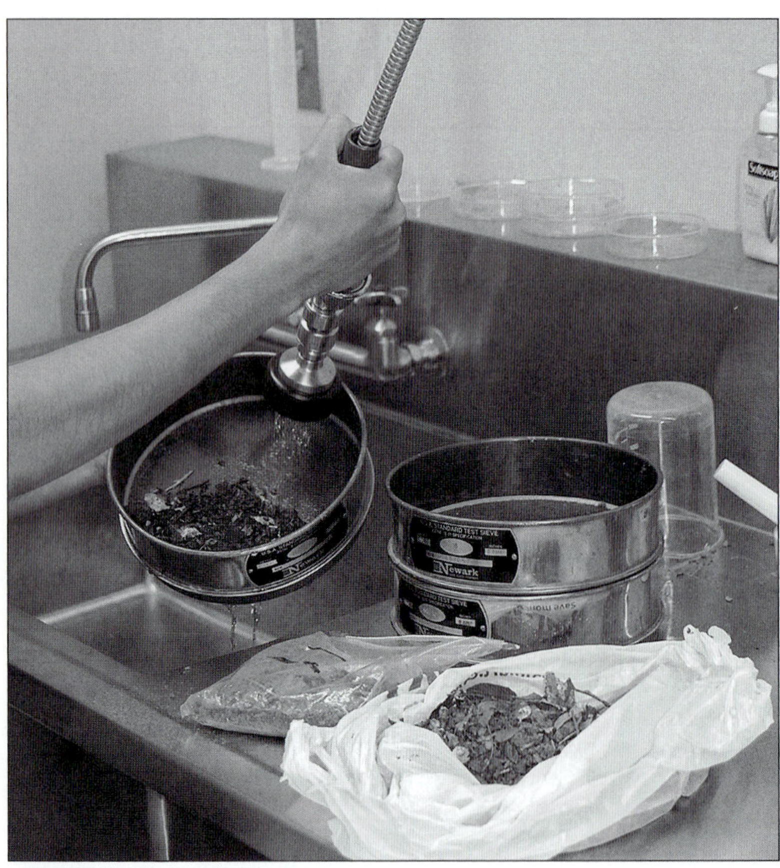

Figure 8.6

Washing a sediment sample through individual or stacked sieves separates soil particles into known size classes. Soil water retention depends as much on diversity of particle sizes as it does on the mean particle size.

6. Rinse the sediment retained on each sieve into clean, pre-weighed, labeled jars. Dry at 105°C for 24 h and weigh the sediment representing each size class. Record the weight of each sediment size class in table 8.1.
7. Calculate for each sediment size class the percent of the total sample weight and record in table 8.1.
8. Repeat steps 1–7 for each transect.

Questions 6

What is the relationship between sediment particle size and stream velocity? ______________________

____

Some large particles (rocks) are apparently too large for the stream to have moved them. How did they get there?

____

____

Some land uses promote erosion. Does the clarity of the water at your sampling site indicate upstream erosion?

____

____

BIOTIC CHARACTERISTICS

As with terrestrial ecosystems, an aquatic system is an amalgam of species constrained by physical factors, biotic factors, and balancing interactions. Assessing a stream's biota typically requires sampling its vascular plants, invertebrates (insects and crustaceans), fish, and **periphyton** (the microfloral community growing on firm substrate).

Procedure 8.7

Assess the primary producer population.

1. Inspect periphyton, detritus, and vascular plant organic matter visible on the sediment or growing from the sediment.
2. Estimate the percent of sediment covered by periphyton in riffles ______ and in pools ______.
3. Estimate the percent of sediment covered by detritus for riffles ______ and for pools ______.
4. Estimate the percent of sediment covered by vascular plants for riffles ______ and for pools ______.

Questions 7

Which segments of the stream have the most extensive periphyton community, pools, or riffles? ______________________

____

What is the relationship between reception of direct sunlight and extent of plant and periphyton coverage?

____

____

Does a vascular plant stand or area of periphyton require continuous, direct sunlight? ______________________

____

Rivers and streams can be divided vertically into the water surface, the water column, and the bottom, or **benthic**, zone. The benthic zone includes the surface of the stream bottom and the porous interior of the substrate through which surface water routinely flows.

Procedure 8.8

Assess the benthic invertebrate population.

1. Examine the parts and dimensions of a Surber sampler (fig. 8.7). Select a representative pool and riffle to sample for invertebrates. The depth should be less than the height of the sampler.
2. Randomly select five sites to take a 1-ft^2 benthic sample within the pool or riffle.
3. For the first site, face upstream and lower the Surber sampler to the stream bottom with the mouth facing upstream. The flowing water should expand the catch net.

TABLE 8.1

SEDIMENT PARTICLE SIZE DISTRIBUTION FOR A MIDCHANNEL SAMPLE FROM A RIFFLE AND A POOL

Pool Transect			Riffle Transect		
Sediment Particle Size Range (mm)	Dry Weight (g)	Percent of Total Sample Dry Weight	Sediment Particle Size Range (mm)	Dry Weight (g)	Percent of Total Sample Dry Weight

Figure 8.7

A Surber sampler depends on a steady water flow to sweep dislodged invertebrates into a trailing net. Be patient and thorough when digging, stirring, and scrubbing the gravel and rocks of a stream bottom.

4. Firmly hold the frame against the substrate. Pick up and hold the large rocks in front of the net and brush their surface thoroughly to dislodge clinging invertebrates. Be sure that the material dislodged from the rocks flows into the Surber net. Put the brushed rocks to the side.
5. After all the large rocks have been brushed, use a trowel to stir the sediment within the square-foot frame. Stir the sediment thoroughly to a depth of 5–10 cm.
6. Lift the sampler while retaining the caught organisms in the net, and repeat steps 3–5 for all five replicate samples.
7. When all five replicates have been taken, empty the pooled samples from the net into a wide-mouth jar. Return the live sample to the lab for sorting. If necessary, preserve the sample with 40% isopropyl alcohol.
8. At the lab, pour the contents of the sample jar from each pool or riffle sampled into large, shallow, white trays. Use forceps to remove the macroinvertebrates (> 3 mm) and sort them in petri dishes by general body shape.
9. Examine figure 8.8 to determine the taxonomic order for each sorted group of insects or the major invertebrate taxon for that group.
10. Use aluminum foil to form a small weighing pan for each taxon, weigh the pan, and add the organisms.
11. Dry the organisms for 24 h at 105°C, weigh the pan with organisms, and subtract the original weight of the pan to determine the dry weight of the organisms.
12. Record in appropriate places on your map the dry weight (g m^{-2}) of invertebrates for each of the riffles and pools sampled.

Questions 8

What adaptations of the invertebrate bodies do you see that might help them live in a fast current? ______________

____

Are riffle invertebrates more diverse than pool invertebrates?

____

____

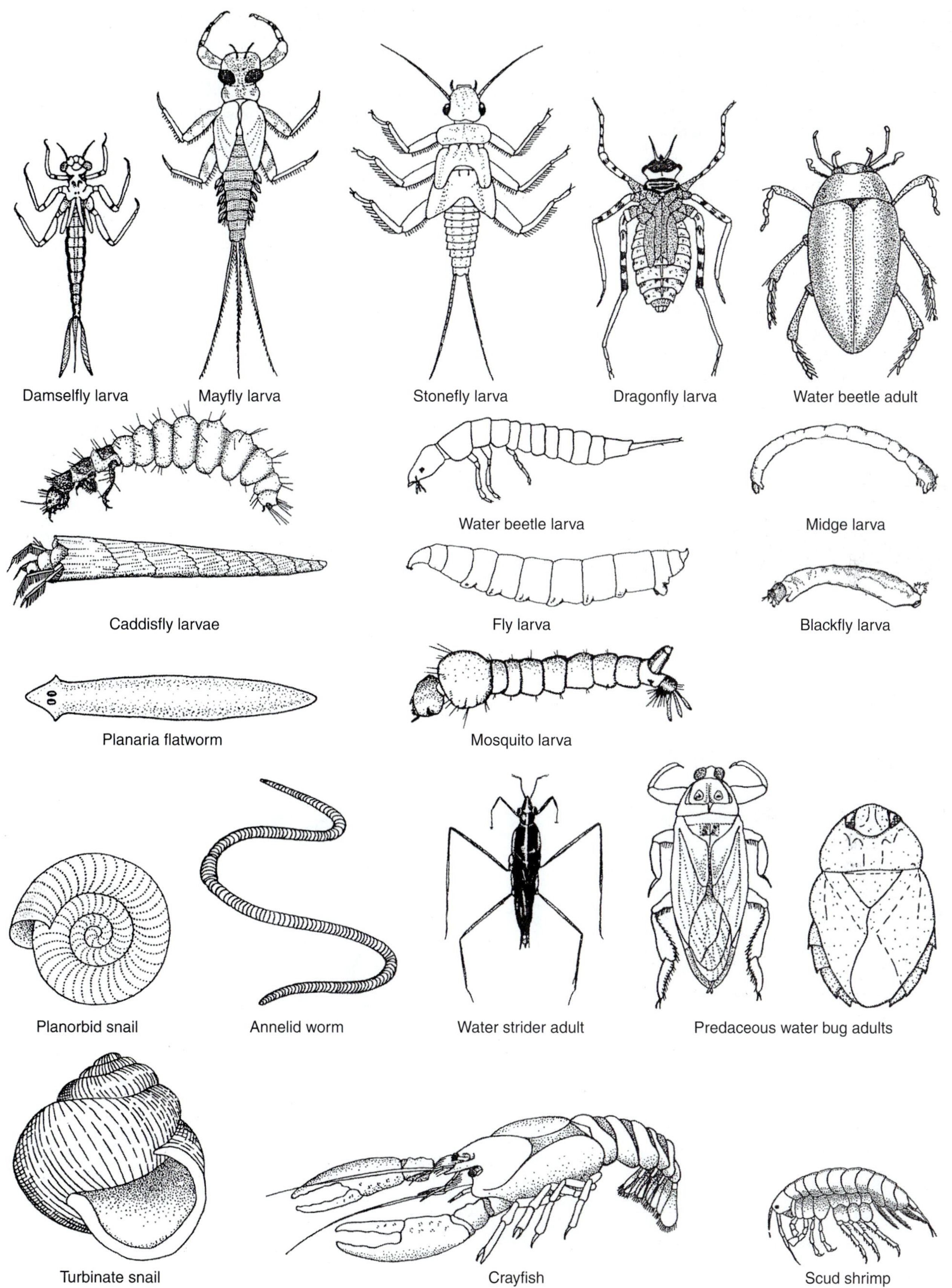

Figure 8.8

Common invertebrates of stream sediments.

Procedure 8.9

Assess and compare the fish populations of riffles versus those of pools.

1. Obtain a 12-ft (or longer) fish seine net. Choose a representative pool and riffle to sample.
2. Your instructor will demonstrate how to effectively use a fish seine (fig. 8.9).
3. Remember these tips:
   a. Sweep the seine upstream rather than downstream.
   b. Hold the seine poles so you push the bottom end in front of you rather than walking backwards and pulling the seine.
   c. Be safe. Move quickly but step carefully and don't overwhelm your seining partner.
   d. While seining with your partner, separate the two poles only about one-half to two-thirds the total length of the seine. For example, the poles of a 12-ft seine should be kept no more than 6–9 ft apart.
   e. Keep the lower, weighted edge of the seine against the sediment. As you move, bump the ends of the poles along the bottom of the sediment. This keeps the seine low in the water column.
   f. To finish a seine haul, sweep the net toward and onto the shoreline rather than lifting the net out of the water while you stand in the stream.
   g. Handle the fish as little as possible. All fish must be returned to the water alive.
4. Have a team of students prepared to count the fish immediately when the net is brought to the shoreline.
5. After each seine haul, count and record the number of each type (species) of fish.
6. For each species, measure the length of 10 representative fish to calculate a mean length.
7. Sketch a lateral view of each species to assess the general shape with special attention to the ratio of length to height.
8. Take two or three seine hauls from each pool and riffle. Combine the data for the pool seine hauls and for the riffle hauls.

Questions 9

Does fish shape correlate with current speed of the immediate environment? ______________________________

____

Figure 8.9

Effective hauling of a seine net depends on using the net as a sack that trails behind the poles rather than a stretched, flat net. Always keep the net moving, and keep its bottom edge dragging on the stream or lake bottom.

What is the relationship between areas of high current velocity and fish diversity? ______________________

____

Which environment—pool or riffle—supports the greatest apparent fish species diversity? ______________________

____

What are the likely features of that micro-environment (pool or riffle) that supports higher fish densities? ______

____

If fish densities are greater in pools (or in riffles), is it a consequence of water velocity? ______________________

____

Questions for Further Thought and Study

1. Would you expect a more rapidly flowing stream to be more diverse? Why?

2. Nutrient and carbon cycling readily occurs within a relatively closed ecosystem. Do streams fit this model of recycling within a closed system? Can you argue that it fits as well as doesn't fit?

3. Water flow can be considered the defining characteristic of a stream. Yet some ecologists say variation in water flow is the defining feature. What might be their logic?

9 Microcommunity Assessment

Objectives

As you complete this lab exercise you will:

1. Collect, examine, and count invertebrates in leaf litter and lichen communities.
2. Pose a research question about an invertebrate microcommunity.
3. State a hypothesis about the ecology of the community.
4. Gather and analyze the data to test your hypothesis.

Small communities often go unnoticed. We usually think of communities that cover areas large enough for us to walk through and support organisms we can easily see. But within these large communities are small, inconspicuous, and thriving **microcommunities** that deserve a closer look (fig. 9.1). A **community** is an association of species that live together. Composition of the community is determined by its species interactions and by characteristics of the environment.

In this lab exercise you will sample two microcommunities: arthropods in woodland leaf litter and invertebrates in lichens. Leaf litter in a woodland includes a variety of small arthropods and members of minor phyla. Lichens might seem like a rather desolate place for a community of microorganisms, but rotifers, nematodes, tardigrades, and protozoa come to life when a lichen is hydrated. After you perfect techniques for collecting, counting, and assessing resident invertebrates in lichen communities and then in leaf litter, you will propose and test hypotheses about their ecology.

Figure 9.1

This woodland community is dense enough to produce much litter and open enough for lichens to prosper.

LICHEN COMMUNITY

Lodged within lichens (fig. 9.2) is a surprising diversity of microscopic (< 1 mm) invertebrates and protozoa that await reviving. The most common invertebrate residents of lichens include nematodes, rotifers, and tardigrades. A single lichen commonly accommodates as many as 5–10 species of tardigrades, rotifers, nematodes, mites, small insect larvae, and various protozoa. Many of these well-adapted animals survive periodic drying in lichens by entering a dormant, metabolic state called **cryptobiosis**. During cryptobiosis, microinvertebrates survive by suspending all but the most vital life functions.

Figure 9.2

This lichen initially appears dry and lifeless, but this combination of algae and fungi is extremely tolerant of desiccation. Addition of water reanimates a variety of microorganisms living on the lichen.

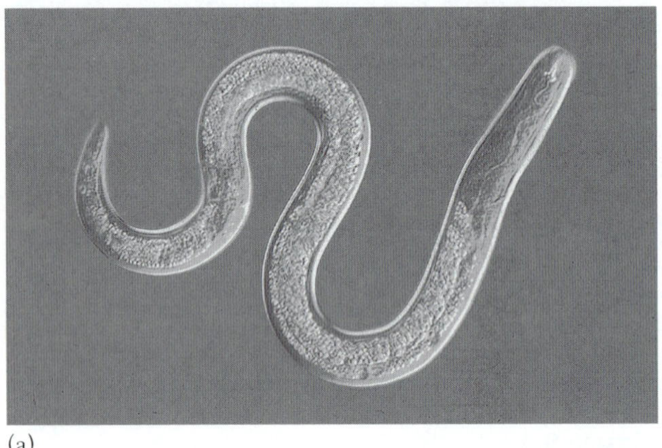

(a)

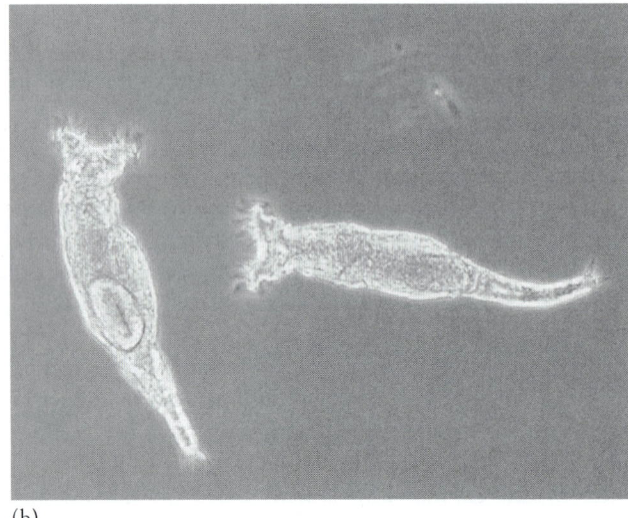

(b)

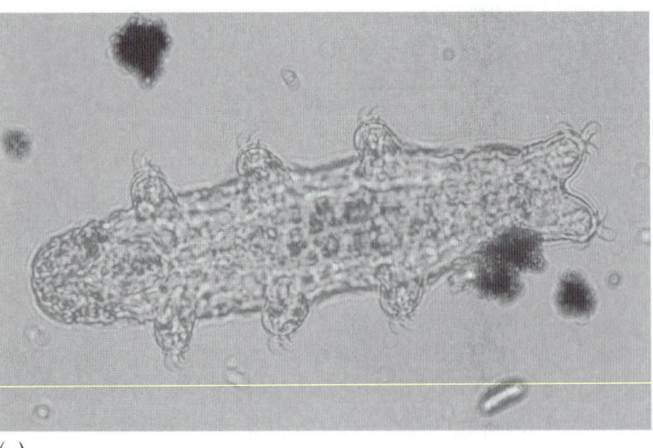

(c)

Figure 9.3

Typical inhabitants of lichens include (a) nematodes; (b) rotifers; and (c) tardigrades.

(a) Nancy Kokalis-Burelle, USDA-ARS; (b) Courtesy of David Mark Welch/MBL; (c) © CH Diagnostic Inc., Fisheries & Oceans, Canada

Tardigrades

Tardigrades (phylum Tardigrada) are unusual and rarely seen even when they are active in a moist environment. These microscopic animals range from 0.1 to 1.0 mm long and live in lichens, mosses, and wet leaf litter. Tardigrades, commonly called "water bears," become active only when surrounded and rehydrated by water. Reanimated tardigrades cling to substrate and search slowly for food. When their surrounding water evaporates, tardigrades eliminate as much as 90% of their body water and assume a desiccated form called a "tun." This loss of body water, called **anhydrobiosis**, leads to a cryptobiotic state in which these organisms can survive for months or even years until reanimated with water.

Nematodes

Nematodes (phylum Nematoda), commonly called roundworms, include 12,000 recognized species. They are remarkably abundant and diverse in marine, freshwater, terrestrial, and parasitic habitats. Most nematodes are microscopic (< 1 mm) and live in soil and sediment. A scoop of fertile soil may contain more than a million nematodes. They are slender, long, and rather featureless worms that feed on detritus and cellular fluids of plants and animals.

Rotifers

Rotifers (phylum Rotifera) are small (< 0.5 mm), bilaterally symmetrical, aquatic animals with a crown of cilia at their heads. Their active cilia filter organic particles from the environment as food. Most of the 2000 species of this interesting phylum live in freshwater, soil, or damp crevices of plants and lichens. Rotifers are a significant member of lake zooplankton communities.

The following procedure is a protocol for sampling lichens and assessing the resident community. This protocol will be part of your experimental design to test a hypothesis that you and your team propose in Procedure 9.2.

Procedure 9.1

Sample a lichen microcommunity.

1. Scrape four replicate lichen samples from tree bark, or cut the lichen and bark away from the branch if the lichen does not release easily. Record where the samples were taken (table 9.1, *Location*).
2. If the lichens are still attached to tree bark, trim the bark so the remainder is covered 100% by lichens. Place the replicate samples in plastic bags labeled as Rep 1, 2, 3, and 4 and return them to the laboratory.
3. For each replicate lichen, trim the edges so it will fit into a petri dish.
4. Determine the area of each lichen.
   a. Outline each lichen on graph paper with lines marking four squares per centimeter (10 squares per in.).
   b. For each replicate lichen, count all squares completely enclosed in its outline. Record the number in table 9.1.
   c. Count all squares the outline subdivides. Record the number in table 9.1.
   d. For each outline, calculate and record in table 9.1 the *Total Squares* by adding the number of complete squares to half the number of subdivided squares.
   e. Divide the *Total Squares* by 16 to convert to *Area of Lichen* (cm^2). Record the areas of each lichen in table 9.1 and the bottom of table 9.3.
5. Label a petri dish for each lichen replicate and fill each dish half full with filtered pond water (or distilled water). Invert each lichen replicate (lichen side down) in its petri dish. Make sure the entire lichen surface is submerged.
6. After 24 h remove the lichen from each dish and use a dissecting microscope to scan the petri dish contents at 50×. Search for moving invertebrates and record the number and species in table 9.2.

Table 9.1

Calculation of the Area of Replicate Lichen Samples

Location ________________

	Number of Completely Enclosed Squares	Number of Divided Squares	Total Squares	Lichen Area (total squares / 16)
Replicate 1				cm^2
Replicate 2				cm^2
Replicate 3				cm^2
Replicate 4				cm^2

7. Use a Pasteur pipet to remove each invertebrate and place it in a labeled vial of 50% ethanol for later study or mounting.
8. Return the lichen to the water.
9. After 24 h more, repeat steps 6–8 and record in table 9.2 the invertebrates found during second 24h.
10. After 24 h more, repeat steps 6–8 and record in table 9.2 the invertebrates found during third 24h.
11. Examine your raw data set in table 9.2. For each replicate and each species, record in table 9.3 the sum of individuals (from table 9.2) collected at all three time intervals (24 h, 48 h, and 72 h).
12. Calculate and record in table 9.3 the density (number cm^{-2}) of each species of invertebrate in each replicate by dividing the number of each invertebrate species by the area of the lichen replicate.
13. Calculate and record in table 9.3 the *Mean Densities* of invertebrates by summing the four replicate densities of each invertebrate species and dividing by 4.
14. Calculate and record in table 9.3 the *Density of Total Invertebrates* as the sum of the mean densities of all invertebrate species.
15. Calculate and record in table 9.3 *Shannon's Diversity* of all invertebrates. See Exercise 12 for steps to calculate diversity.

Questions 1

What group of invertebrates was the most abundant in your lichen samples? Least abundant? ____________________

____

Was your measure of lichen surface area accurate? Is there a better way? ____________________

____

Lichen communities, like all organisms, encounter an array of physical, chemical, and biological conditions that vary from one environment to another. Good research involves identifying ecological conditions (variables) likely to impact the community and then testing the effects of those variables.

Table 9.2

Data for microinvertebrates extracted from four replicate lichen samples

	Invertebrates during 1st 24h				Invertebrates during 2nd 24h				Invertebrates during 3rd 24h			
	Rep 1	Rep 2	Rep 3	Rep 4	Rep 1	Rep 2	Rep 3	Rep 4	Rep 1	Rep 2	Rep 3	Rep 4
Species A												
Species B												
Species C												
Species D												
Species E												

Table 9.3

Data sheet for calculation and analysis of density and diversity of microinvertebrates extracted from a lichen sample

	Total Invertebrates Recovered after 72 h (summed from table 9.2)				Density of Invertebrates (number cm^{-2} lichen)				Mean Density (sum of reps 1–4 / 4)	Shannon Wiener Diversity (H')	
	Rep 1	Rep 2	Rep 3	Rep 4	Rep 1	Rep 2	Rep 3	Rep 4			
Species A										$(pA)(\log_e pA)$	
Species B										$(pB)(\log_e pB)$	
Species C										$(pC)(\log_e pC)$	
Species D										$(pD)(\log_e pD)$	
Species E										$(pE)(\log_e pE)$	
Lichen area (cm^2)					Density of Total Invertebrates =				____		H' ____

9–4

Procedure 9.2

Design and test a hypothesis involving factors that affect the density and diversity of microinvertebrates in a lichen community.

1. Review Exercises 1 and 2.
2. After discussion with your lab group, list 5–10 ecological factors likely to influence the density and diversity of invertebrates in a lichen microcommunity.
   1. ______________________________
   2. ______________________________
   3. ______________________________
   4. ______________________________
   5. ______________________________
3. Identify the ecological variable you want to investigate. ______________________________

   ____
4. Pose a general research question that relates this variable to the density and/or diversity within contrasting lichen communities. State your question here: ______________

   ____

   ____
5. Pose a testable hypothesis that will provide at least a partial answer to your question. State that hypothesis here: ______________________________

   ____

   ____
6. Discuss with your lab group and instructor an experimental design to test your hypothesis.
7. Design your raw data sheet and analysis sheet (similar to tables 9.2 and 9.3), and conduct your experiment.

Questions 2

Did you accept or reject your hypothesis? ______________

What was the answer to your question? ______________

____

Was your experimental design adequate to answer your question fully? ______________________________

____

How would you improve your experimental design? ______

____

____

Good research usually leads to further questions. How would you expand your research to better answer your question?

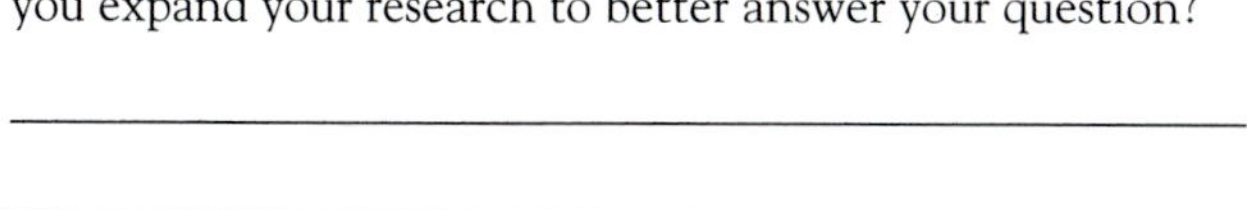

____

Figure 9.4

Leaf litter is a remarkably hospitable habitat for large and small invertebrates. Leaf litter is moist, insulating from the heat of direct sunlight, and rich in organic detritus as food for decomposers.

LEAF LITTER ARTHROPOD COMMUNITIES

Decaying leaf litter with abundant organic material supports a microcommunity of invertebrates (fig. 9.4). This web of invertebrates is supported by **detritus** (decomposing organic matter) and by bacteria and fungi feeding on the detritus. Invertebrates, especially small arthropods, are abundant grazers on the detritus and microbial decomposers and may number into thousands per square meter. As you might guess, the density and diversity of arthropods in leaf litter and the upper surface of the soil vary significantly with temperature, moisture, and organic input.

Ecologists extract small arthropods from a sample of leaf litter with a **Berlese funnel** (fig. 9.5). A sample of leaf litter is placed on a screen below a light bulb. The bulb's light and warmth causes the arthropods to move down through the sample, fall through the screen, and slide down the funnel into a collecting vial with preservative.

Procedure 9.3

Collect leaf litter arthropods from a woodland forest floor.

1. Assemble four Berlese funnels, and locate a woodland sampling site with moist and abundant leaf litter (> 2.5 cm thick).

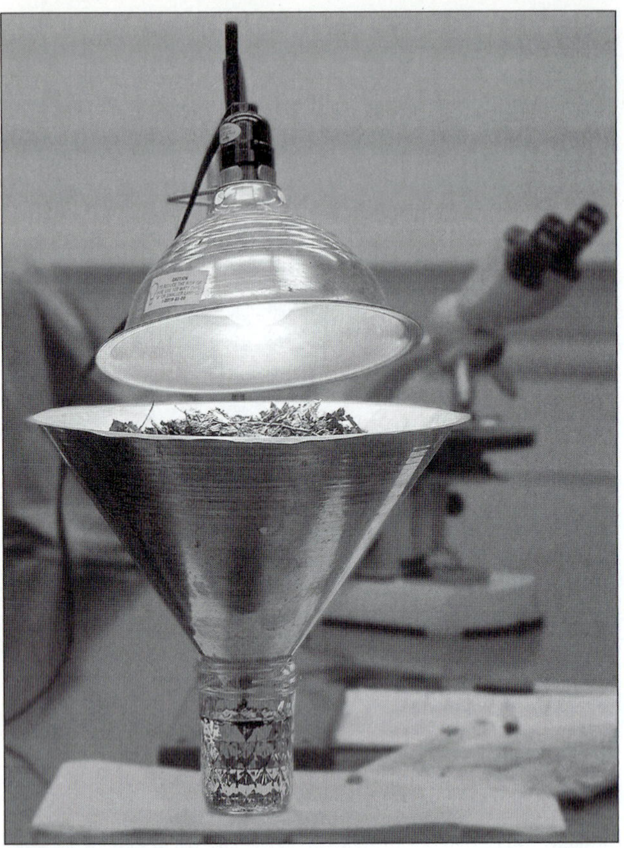

Figure 9.5

A Berlese funnel uses heat from a light bulb to drive invertebrates toward the bottom of a leaf litter sample where they fall through a screen into a preserving solution.

2. Collect in plastic bags four replicate samples of leaf litter from patches (about 200 cm^2) of woodland forest floor. Collect deep enough to include any loose soil to a depth of about 1 cm. A reasonable sample size (area) depends on the amount your Berlese funnel can accommodate.
3. Return to the lab and put the samples on the screens in the funnels. Some material may fall through immediately. Gather the fallen material and put it on top of the leaf litter so it won't fall through again.
4. Place a collecting vial with 50% ethanol under each funnel spout, and turn on the light bulb. Do not overheat the samples.
5. Allow the invertebrates to collect in the vials for 12 h.
6. Replace the sample vial for each funnel every 12 h for 36 total h. If the sample is particularly wet, you may need to flip the sample on the screen rather than let the upper surface burn before the sample dries to the bottom.
7. Sort the collected arthropods into major groups (fig. 9.6). Record in table 9.4 the number of each species collected.
8. Examine your raw data set in table 9.4. For each replicate and each invertebrate species, record in table 9.5 the *Total Arthropods Recovered* at 12 h, 24 h, and 36 h.
9. Calculate and record in table 9.5 the *Density of Arthropods* for each species in each replicate by dividing the number collected of that species by the sample area of the replicate.
10. Calculate the mean densities of arthropods by summing the four replicate densities of each arthropod species and dividing by four.
11. Calculate and record in table 9.5 the *Density of Total Arthropods* as the sum of the mean densities of all invertebrate species.
12. Calculate and record in table 9.5 *Shannon's Diversity* of all invertebrates. See Exercise 12 for steps to calculate diversity.

Table 9.4

Number and Species of Arthropods Collected from Four Replicate Samples of Leaf Litter

	Arthropods during 1st 12h				Arthropods during 2nd 12h				Arthropods during 3rd 12h			
	Rep 1	Rep 2	Rep 3	Rep 4	Rep 1	Rep 2	Rep 3	Rep 4	Rep 1	Rep 2	Rep 3	Rep 4
Species A												
Species B												
Species C												
Species D												
Species E												
Species F												
Species G												

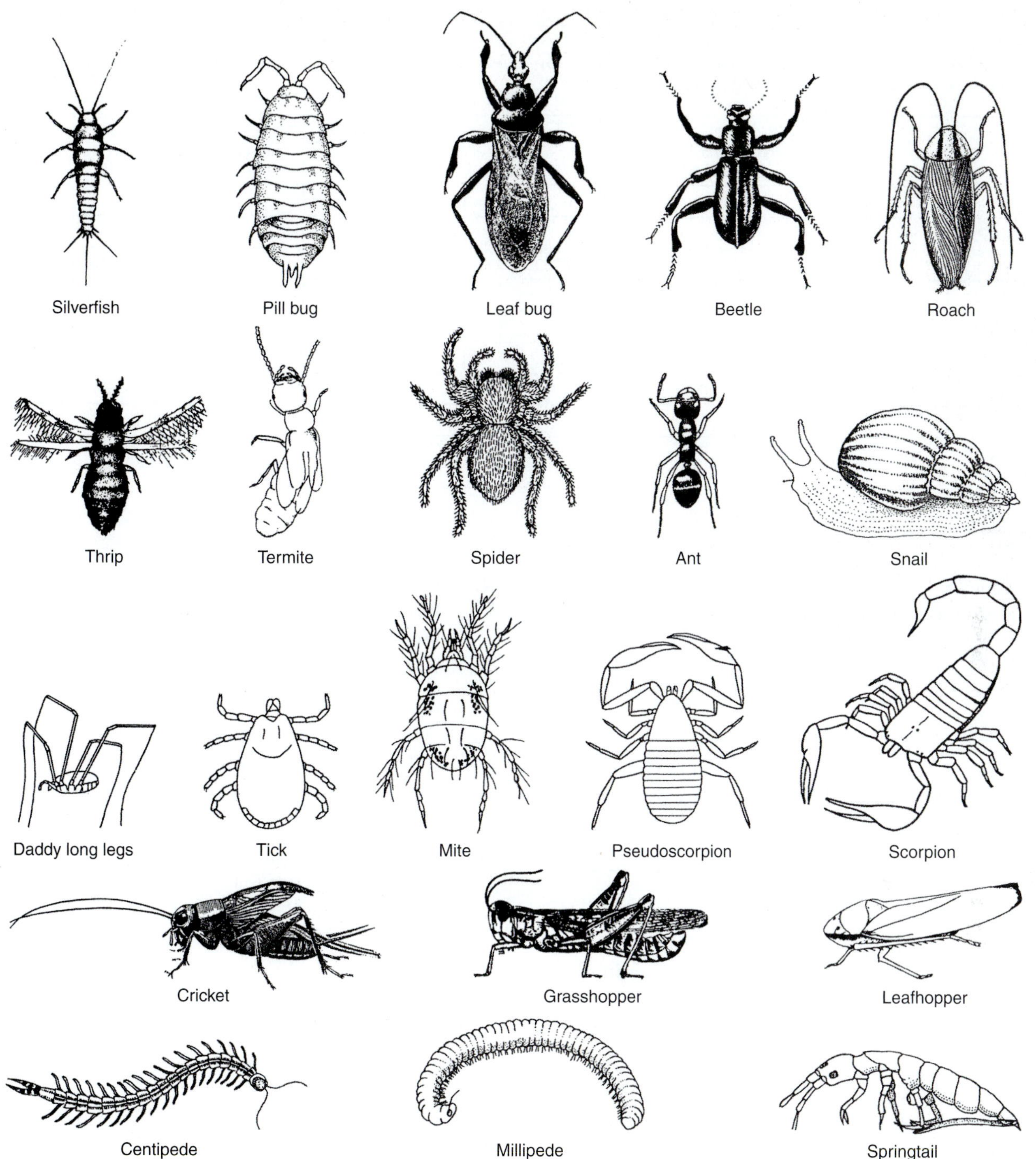

Figure 9.6

Representative soil invertebrates.

Table 9.5

Data sheet for calculation and analysis of density and diversity of arthropods extracted from a leaf litter sample

	Total Arthropods Recovered (summed from table 9.1)				Density of Arthropods (number cm^{-1} sample)				Mean Density (sum of reps 1–4 / 4)	Shannon Wiener Diversity (H')	
	Rep 1	Rep 2	Rep 3	Rep 4	Rep 1	Rep 2	Rep 3	Rep 4			
Species A										$(pA)(\log_e pA)$	
Species B										$(pB)(\log_e pB)$	
Species C										$(pC)(\log_e pC)$	
Species D										$(pD)(\log_e pD)$	
Species E										$(pE)(\log_e pE)$	
Species F										$(pF)(\log_e pF)$	
Species G										$(pG)(\log_e pG)$	
Sample area (cm^2)					Density of Total Arthropods =				_____		H' _____

Questions 3

What arthropods were most common in leaf litter? ______

____

Least common? ____

____

Is a Berlese funnel a good device to extract leaf litter arthropods? Would it work well for all microorganisms? Why or why not? ____

____

Procedure 9.4

Design and test a hypothesis concerning factors affecting the density and diversity of arthropods in a leaf litter community.

1. Review Exercises 1 and 2.
2. After discussion with your lab group, list 5–10 ecological variables likely to influence the density and diversity of arthropods in a leaf litter community. List these factors in the following space.
   1. ____
   2. ____
   3. ____
   4. ____
   5. ____
3. Identify the ecological variable you want to investigate. ____

   ____

   ____
4. Pose a general research question that relates this variable to the density and/or diversity among leaf litter arthropods from contrasting communities. State your question here. ____

   ____
5. Pose a testable hypothesis that will provide at least a partial answer to your question. State that hypothesis here.____

   ____
6. Discuss with your lab group and instructor an experimental design to test your hypothesis.
7. Design your raw data sheet and analysis sheet (similar to tables 9.1 and 9.2), and conduct your experiment.

Questions 4

Did you accept or reject your hypothesis? ______________

What was the answer to your question? ______________

____

Was your experimental design adequate to answer your question fully? ____

____

How would you improve your experimental design? ______

____

____

Good research usually leads to further questions. How would you expand your research to better answer your question?

____

____

Questions for Further Thought and Study

1. What are the major ecological factors that influence the density of leaf litter arthropods?

2. What major ecological factors would influence the invertebrate community in a lichen?

3. For which other environments would lichen invertebrates be well adapted?

4. What other microcommunities occur in a forest ecosystem? Are microcommunities self-contained? How so?

5. What role did sample size play in testing your hypotheses?

Sampling a Plant Community

10

Objectives

As you complete this lab exercise you will:

1. Survey plant species richness and habitat gradients in a local environment.
2. Assess herb, shrub, and tree densities with methods of quadrats, strip transects, line transects, and line intercepts.
3. Calculate densities, frequencies, and relative importances of herb, shrub, and tree communities.
4. Assess stratification of plant communities along a gradient.

Communities would be much easier to study if plants were regularly distributed in space and time. But they aren't. Plants don't grow in uniform, spatial patterns. Put simply, plant distributions vary because the environment is patchy, and the distribution of all organisms ultimately depends on patchy resources, both biotic and abiotic. This patchiness is so pervasive in natural communities that ecologists need rigorous sampling techniques and large sample sizes to overcome natural variation (fig. 10.1).

The fundamental goals of sampling a plant community are to determine (1) which plants live there; (2) how many plants occur per square meter; and (3) which species dominate the local environment. These aren't easy to determine. Sample size, shape, and number are all critical for good results. The more large and uniform samples you take, the better the estimate, but the greater the cost and effort. The best sampling designs balance sampling effort with accuracy.

Question 1

Would you expect animal distributions to be as patchy as those of plants? Why or why not?____________________

____

In this lab exercise you will use a variety of sampling methods and learn the best applications and effort required for each method. Most sampling designs involve randomly distributing as many sampling units (quadrats and transects) over the study area as possible and counting the plants in each sample. Randomization ensures that your data will meet assumptions of statistical analyses. Always remember that these quantitative procedures are further enhanced by your initial observations of the species present and of variations in soil, moisture, disturbance, drainage, etc. Raw data is gathered objectively, but the quality of the questions asked and the overall sampling design require walking, observing, and examining the environment through the eyes of an ecologist.

Figure 10.1

All of the plants in this 1-m^2 quadrat can be counted in just minutes. Counts from multiple, randomly placed quadrats of a known area represent samples used to estimate plant densities in a much larger area.

Procedure 10.1

Assess by observation the species richness, habitat gradients, and boundaries of a plant community.

1. Review Exercise 7. Decide which observations relate to the environment under study.
2. Discuss with your instructor the general location and boundaries of a study site(s).
3. Divide into teams to qualitatively (by observation) compile a list of tree species, shrub species, and herb species. One team should also survey boundaries and gradients of abiotic factors (soil, moisture, inclines, etc.).

TABLE 10.1

QUALITATIVE LIST OF TREE SPECIES, SHRUB SPECIES, AND HERB SPECIES

Trees		Shrubs		Herbs	
Species	Identifying Characteristic	Species	Identifying Characteristic	Species	Identifying Characteristic

Observations of abiotic factors and gradients:

4. Each team should record in table 10.1 the species list for its assigned taxon (trees, shrubs, herbs).
5. The team assessing abiotic factors should record its observations in table 10.1 and prepare sketched maps of each relevant abiotic factor.
6. Teams should consult with each other and define the boundaries of major tree-, shrub-, and herb-dominated areas of the broader environment. Sketch these boundaries.

Questions 2

Practiced observation is a powerful technique. How would you improve your powers of observation and the accuracy of your species list? ____________________

____

What are the major drawbacks to relying solely on observation to assess a plant community? ____________________

____

QUADRATS AS SAMPLING UNITS FOR HERBS

Many methods measure numbers, densities, and distributions of plants in terrestrial communities. Probably the most widely used method is **quadrat sampling.** Uniform quadrats of a known area (sometimes called **plots**) are randomly distributed in a habitat and the organisms in each quadrat are counted. For quadrat sampling, the **frequency** of a species is the percent of all sampling units that have at least one individual. Because a quadrat is a known area, the results for one species, or for all species, can also be expressed as **absolute density**. Density is the number of organisms per unit area. Occurrence of a species can also be expressed as **relative density**, which is the percentage of the total number of individuals represented by that species.

Choosing the *size* of the quadrat depends on the density of plants being sampled. Typical quadrats are squares of 1 square meter or more. They should be large enough to frequently contain five or more individuals, but small enough for you to count all individuals in a reasonable time. Herbs, shrubs, and trees are sampled well by quadrats 1 m^2, 4–10 m^2, and 100–500 m^2, respectively.

The term *quadrat* specifically refers to a four-sided rectangle, but quadrats can be any shape, including a circle (fig. 10.2). *Shape* is usually determined by ease of layout. Some ecologists prefer to use a circular quadrat to minimize edge length and to eliminate problems in determining if a plant is inside or outside the edge. Circular quadrats are easily laid out by pushing a large metal pin (nail) into the ground, tying a string to the pin so the string can slip around the pin, marking the string at 56.4 cm (radius) from the pin, and rotating the string around the pin to delineate a 1-m^2 circle. A radius of 79.8 cm encompasses a 2-m^2 quadrat.

The *number* of quadrat samples should be as large as reasonable, and depends on the nature of the plants being studied and the effort needed to count each quadrat. Thirty quadrats usually produce reliable results.

Figure 10.2

Circular plots are good sampling units and can be easier to define than a square plot. A radial string is rotated around a center point to define the perimeter of a circle of known area. The students are counting plants inside the circle.

Procedure 10.2

Assess herb density using the quadrat method.

1. Refer to the results of Procedure 10.1, step 6 and determine the boundaries of the area to sample herbs. For sampling purposes, herbs are defined as nonwoody plants less than 40 cm high.
2. Consult with your instructor and choose quadrat size (1 m^2 recommended), shape (square or circle), and number of quadrats for sampling.
3. Quadrats must be placed randomly. Using the following steps, establish an imaginary grid with numbered positions over the area being sampled. To establish the grid:
   a. Establish and mark the ends of a baseline along one edge of the area being sampled. Make the baseline as long as the longest edge of the area.
   b. Mark the baseline at 1-m intervals.
   c. At one end of the baseline, establish a perpendicular line long enough for a grid of positions defined by the two lines to overlay the entire area being studied. Mark the perpendicular line at 1-m intervals. Points (1-m intervals) along the baseline represent columns. Points (1-m intervals) along the perpendicular line represent rows (fig. 10.3).

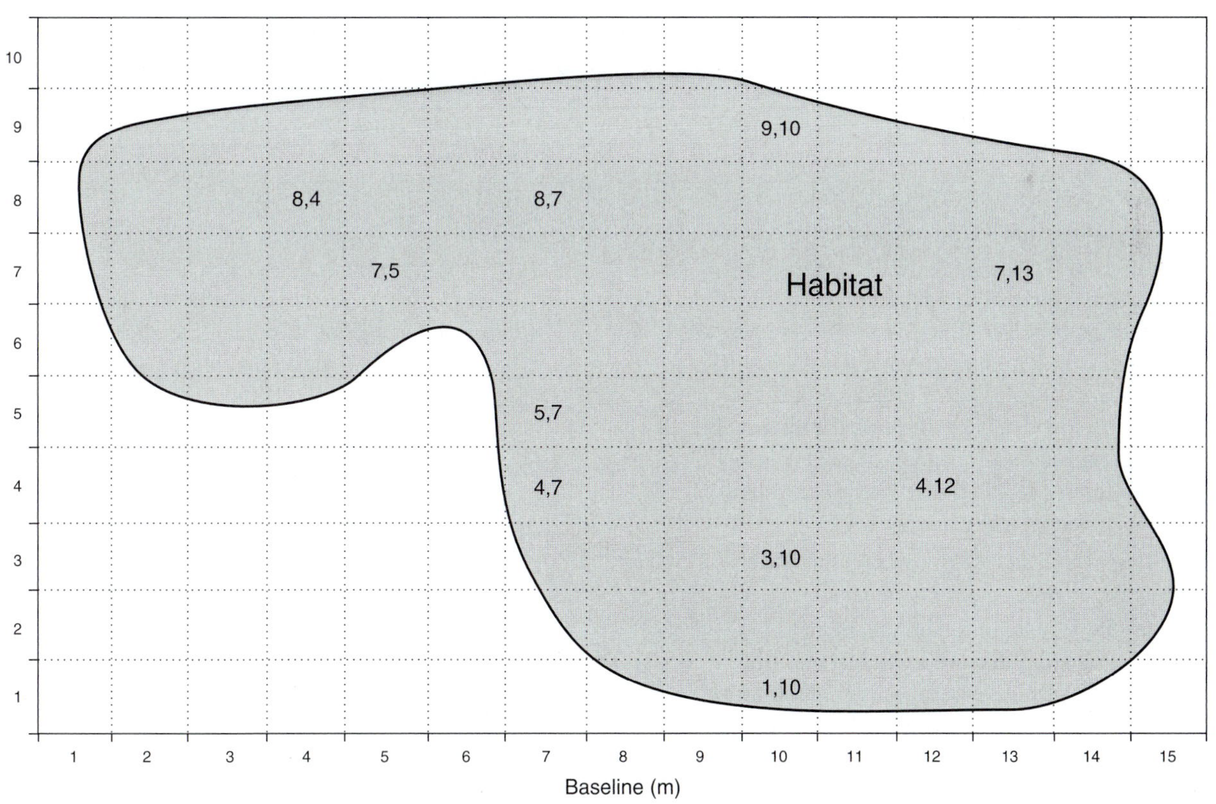

Figure 10.3

A baseline and perpendicular line position a grid over an irregular area of a habitat to be sampled. Mark the lines in 1-m intervals. Then choose row and column numbers randomly to locate 10 replicate quadrats for sampling.

Table 10.2

The number of each herb species found in 20 randomly positioned quadrats

Species	1	2	3	4	5	6	7	8	9	10	11	12	13	14	15	16	17	18	19	20	# Quadrats With at Least One Individual	Total Number of Individuals $Species_i$
	Quadrat														One quadrat = ______ m^2							
Total herbs																						

Table 10.3

Densities of herbs sampled with quadrat sampling units

Species	Frequency (%)	Absolute Density (num m^{-2})	Relative Density (%)
Total Herbs			100%

d. To position a quadrat in a random box of the grid, use a random number table to select a column. Then use a random number table to select a row. The intersection of the row and column is the position of the quadrat. Repeat random column and row selections for each quadrat needed. Record the coordinates (column, row) for all expected quadrats before you begin sampling.

4. Familiarize your team with the species list of herbs in table 10.2.
5. Place the pin (or quadrat square) at the first coordinates. Count and record in table 10.2 the number of each species of herb whose basal (ground-level) stems lie within or on the quadrat boundaries. Sum these numbers and record in table 10.2 the total herbs in the quadrat.
6. Repeat step 5 for each set of coordinates (i.e., each quadrat).
7. For each $species_i$ and for total herbs, calculate and record in table 10.2 the number of quadrats that included at least one individual of that species.
8. Calculate and record in table 10.2 for each species the total individuals counted in all quadrats.
9. Calculate and record in table 10.3 the frequency, absolute density, and relative density for each $species_i$.

$frequency_{species_i}$ = (100 × number of sampling quadrats with one or more individuals of $species_i$) / total number of sampling quadrats

absolute $density_{species_i}$ = total number of $individuals_{species_i}$ / total area sampled

relative $density_{species_i}$ = (100 × total number of $individuals_{species_i}$) / total number of individuals of all species

Questions 3

Did the random placement of your quadrats adequately sample the community? Why or why not? ________________

____

The densities of herb species are rarely even. How many herb species include 90% of the total herb plants? ________

____

Consider absolute density and relative density. Is one more informative than the other? ________________

____

Are the species with highest frequency also the ones with highest density? ________________

____

Which of the three parameters in table 10.3 requires the least effort to gather the necessary data? Why? ________

____

Which season would be best for sampling a plant community? What difference might it make? ________________

____

LINE TRANSECTS AS SAMPLING UNITS FOR HERBS

Another common sampling technique is the **line intercept method**. In this method, transects, or lines, are established and randomly laid out within the community. Organisms that touch or "intercept" the transect are counted and measured (fig. 10.4). Calculations with these data reveal the relative abundances, frequencies, and distributions of the plant species that compose the community.

A disadvantage of the line intercept method is that it does not measure ground surface area. Therefore, you cannot calculate absolute density. You can, however, calculate measures of *relative* density among plant species. This reveals which plants are more dense than others, but not their absolute densities per unit area. Oftentimes ecologists are just as interested in relative density as they are in absolute density.

Space is a precious resource for plants, especially the area that a plant occupies with sunlight. **Coverage** is a measure of aerial space a plant occupies. A species' **relative coverage** is the percent of total plant coverage represented by that species.

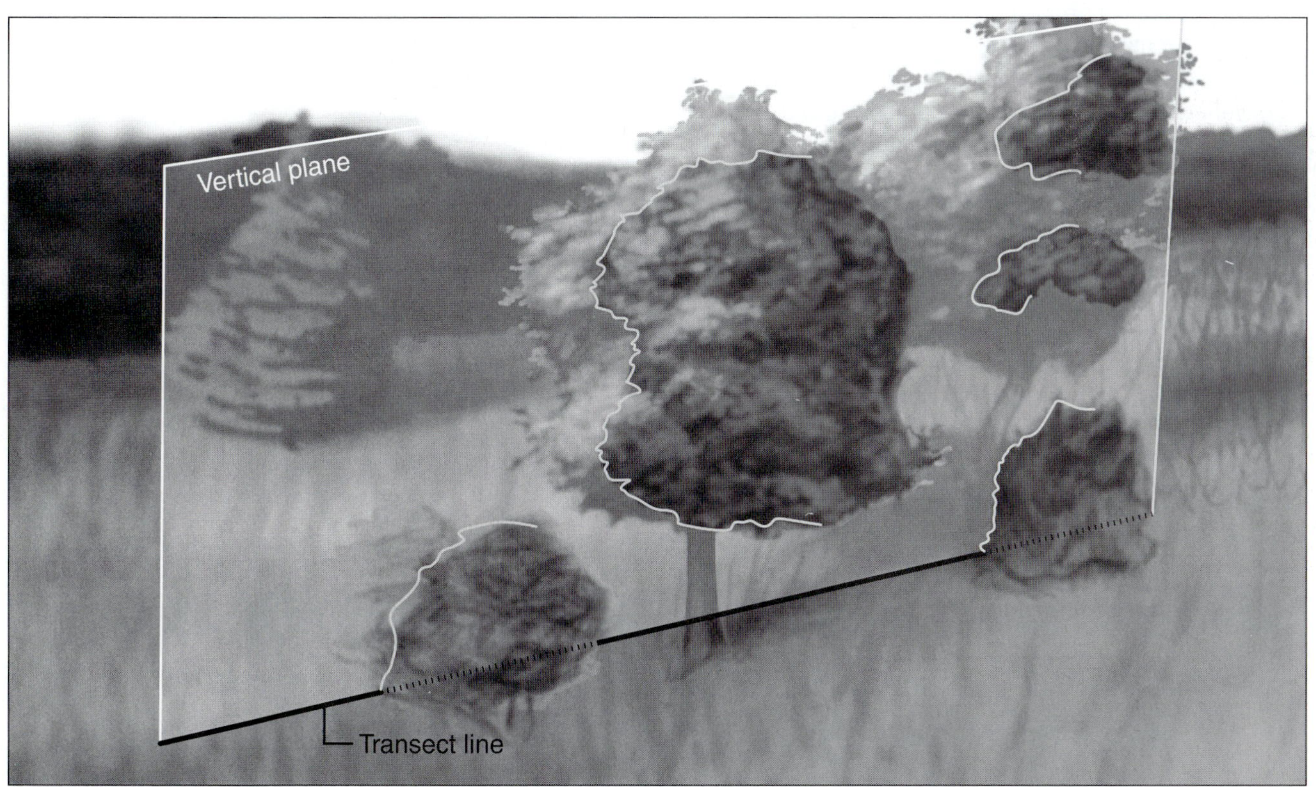

Figure 10.4

A transect includes a line (or tape measure) on the ground plus the imaginary vertical plane extending above the line. Plants rooted to the side of the line but extending into the vertical plane are counted as well as plants touching the ground line are all part of the sample.

Question 4
Is coverage directly related to density? Why or why not?

____

____

Experience with plant sampling quickly reveals that no single parameter fully describes a plant community. No single value measures "success." Therefore, ecologists usually measure a variety of parameters to understand the community. One combination of relevant parameters is an **importance value**. This value combines the data for two or more variables such as density, coverage, frequency and others. In the next procedure you will calculate importance value as a combination of relative density and relative coverage.

Procedure 10.3

Assess the frequencies, relative densities, and importance values of herb species by using line transect and line intercept methods.

1. Refer to the results of Procedure 10.1 and determine the boundaries of a suitable area to sample herbs. For sampling purposes, herbs are defined as nonwoody plants less than 40 cm high.
2. Consult with your instructor and choose a suitable transect length (5 m recommended) and a suitable number of transects (one per team recommended).
3. For your team's transect, obtain a 5-m measuring tape (or measured rope) and a notepad.
4. Your transect must be randomly placed. Use the following steps to select random starting and ending points for your transects on an imaginary grid with numbered positions encompassing the area being sampled.
   a. Establish and mark the ends of a baseline along one edge of the area being sampled. Make the baseline as long as the longest edge of the area.
   b. Mark the baseline at 1-m intervals.
   c. At one end of the baseline, establish a perpendicular line long enough for a grid of positions defined by the two lines to overlay the entire area being studied.
   d. Mark the perpendicular line at 1-m intervals. Points (1-m intervals) along the perpendicular line represent rows. Points (1-m intervals) along the baseline represent columns.
   e. To randomly position a transect, use a random number table to select a column. Then use a random number table to select a row. The intersection of the row and column is the starting point of the transect.
   f. Select a random number (0–360) to determine the compass direction to extend the 5-m transect from the starting point.
5. Familiarize your team with the species list of herbs in table 10.4.
6. For your first transect, stretch the 5-m measuring tape on the ground from your random set of coordinates (step 4).
7. Divide the transect into five 1-m intervals.
8. Along the first interval, count plants that touch or underlie the line. For each individual plant, record in table 10.4 the species and the length (cm) of the line that the plant intercepts. Similarly, for plants that overhang the line, record in table 10.4 the length of the imaginary vertical plane from the line that the plant intercepts. If necessary, also record as *Species Bare* any uncovered (bare) lengths of the interval. Table 10.4 accommodates one transect with data for seven individuals of each of three species in each 1-m interval.
9. When all plants from all five 1-m intervals of the 5-m transect have been recorded, summarize your raw data in table 10.5.
10. Check that each of the other teams has completed data collection for its transect.
11. Use the data in table 10.5 to calculate the following four parameters for each species. Record your results in table 10.6.

    $\text{frequency}_{\text{species}_i}$ = (100 × number of sampling intervals with one or more individuals of species_i) / total number of sampling intervals

    $\text{relative density}_{\text{species}_i}$ = (100 × total number of individuals of species_i) / total number of individuals of all species

    $\text{relative coverage}_{\text{species}_i}$ = (100 × total intercept length by species_i) / total length intercepted by all plants

    $\text{importance value}_{\text{species}_i}$ = ($\text{relative density}_{\text{species}_i}$ + $\text{relative coverage}_{\text{species}_i}$) / 2

12. Your instructor may ask you to compare or combine your data with those from other transects.

Table 10.4

Raw data for herbs intercepting a transect at each interval

Extend this table if necessary to accommodate more species.

	Interval 1		Interval 2		Interval 3		Interval 4		Interval 5	
	Individual	Intercept (m)	Individual	Intercept (m)	Individual	Intercept (m)	Individual	Intercept (m)	Individual	Intercept (m)
Species:	1st		1st		1st		1st		1st	
	2nd		2nd		2nd		2nd		2nd	
	3rd		3rd		3rd		3rd		3rd	
	4th		4th		4th		4th		4th	
	5th		5th		5th		5th		5th	
	6th		6th		6th		6th		6th	
	7th		7th		7th		7th		7th	
Species:	1st		1st		1st		1st		1st	
	2nd		2nd		2nd		2nd		2nd	
	3rd		3rd		3rd		3rd		3rd	
	4th		4th		4th		4th		4th	
	5th		5th		5th		5th		5th	
	6th		6th		6th		6th		6th	
	7th		7th		7th		7th		7th	
Species:	1st		1st		1st		1st		1st	
	2nd		2nd		2nd		2nd		2nd	
	3rd		3rd		3rd		3rd		3rd	
	4th		4th		4th		4th		4th	
	5th		5th		5th		5th		5th	
	6th		6th		6th		6th		6th	
	7th		7th		7th		7th		7th	
Species: bare	1st		1st		1st		1st		1st	
	2nd		2nd		2nd		2nd		2nd	
	3rd		3rd		3rd		3rd		3rd	
	4th		4th		4th		4th		4th	
	5th		5th		5th		5th		5th	
	6th		6th		6th		6th		6th	
	7th		7th		7th		7th		7th	

Table 10.5

Summary of raw data for herb species intercepting a transect

Transect length = _________ Number of intervals = _________

$Species_i$	Number of Intervals with at Least One Individual	Total Number of Individuals of $Species_i$	Total Intercept Length by $Species_i$
		Total number individuals of all species =	Total intercept length by all plants =

Table 10.6

Relative values of each herb species in a selected community using parameters of the line-intercept method

Species	Frequency	Relative Density	Relative Coverage	Importance Value

Questions 5

Are there any "wrong" places to locate a transect? Why or why not? ________________________________

____

What herb species had the highest frequency? _______ Relative density? _______ Relative coverage? _______ Importance? _______

What is the meaning of an importance value? _________

____

Why would we calculate this in addition to density, coverage, and frequency? ________________________________

____

Did your line intercept sampling of herbs reveal any species you missed with your observations in Procedure 10.1?

____

____

STRIP TRANSECTS AS A SAMPLING UNIT FOR SHRUBS

Ecologists frequently extend a transect to include a strip of area on each side of the transect line. In this way a strip transect can be treated as a long, narrow, rectangular quadrat. Strip transects are especially good for measuring density of plants such as shrubs that are farther apart than herbs. For sampling purposes, shrubs are defined as woody-stemmed plants over 40 cm high with stems < 2.5 cm diameter at 1.5 m above the ground. Stem diameter at 1.5 m above the ground is called **diameter at breast height (DBH)**. For sampling purposes, trees are woody-stemmed plants typically having a single main stem and a DBH > 2.5 cm.

Procedure 10.4

Use the strip transect method to measure the density of shrubs.

1. Familiarize your team with the species list of shrubs in table 10.1.
2. Refer to Procedure 10.1 and determine the boundaries of a suitable area to sample shrubs.
3. Consult with your instructor and choose a suitable strip transect length and width (20 m × 2 m recommended) and a suitable number of transects (one per team recommended).
4. For your team's transect, obtain a 20-m measuring tape (or measured rope), meter stick, and a notepad.
5. Your transect must be randomly placed. Follow step 4 of Procedure 10.3 to establish the coordinates of randomly placed strip transects.
6. For your transect, stretch the 20-m measuring tape on the ground to establish a transect at your coordinates. Divide the transect length into 1-m intervals.
7. As you move long the transect, use the meter stick perpendicularly on each side of the line to judge which shrubs are included in the strip transect.
8. Count and record in table 10.7 the number of each species of shrub whose basal (ground-level) stems lie at least partially within the area of each interval of the strip transect boundaries.
9. For each species and for total shrubs, calculate and record in table 10.7 the number of intervals with at least one individual of that species.
10. Calculate and record for each species in table 10.7 the total number of individuals counted in all intervals.
11. Use the data in table 10.7 to calculate the following three parameters for each species. Record your results in table 10.8.

    $\text{frequency}_{\text{species}_i}$ = (100 × number of sampling intervals with one or more species_i) / total number of sampling intervals

    $\text{absolute density}_{\text{species}_i}$ = (total number of $\text{individuals}_{\text{species}_i}$) / total area sampled

    $\text{relative density}_{\text{species}_i}$ = (100 × total number of $\text{individuals}_{\text{species}_i}$) / total number of individuals of all species

12. Your instructor may ask you to compare or combine your data with those from other transects.

Questions 6

What shrub species had the highest density? __________

____

Did all three measures of occurrence in table 10.8 portray the same species composition of the shrub community?

____

____

Which parameter in table 10.8 reveals the most evenly distributed shrubs?__________

____

Most unevenly distributed? __________

____

TABLE 10.7

THE NUMBER OF EACH SHRUB SPECIES FOUND IN A RANDOMLY POSITIONED STRIP TRANSECT

	Interval																				One Interval = ______ m²	
Species	1	2	3	4	5	6	7	8	9	10	11	12	13	14	15	16	17	18	19	20	# Intervals with at Least One Individual	Total Number of Individuals in All Intervals
Total shrubs																						

TABLE 10.8

DENSITIES OF SHRUBS SAMPLED WITH STRIP TRANSECT SAMPLING UNITS

Species	Frequency (%)	Absolute Density (num m^{-2})	Relative Density (%)
Total Shrubs			100%

LINE TRANSECTS AS SAMPLING UNITS FOR TREES

Trees are large and spaced relatively far apart. This distribution calls for a method that covers as much territory as possible without extraordinary effort to count all the trees. A line intercept procedure works well, especially if the transect is long and only trees are counted. The protocol for assessing trees is the same as for herbs described in Procedure 10.3, except the transects are longer.

Procedure 10.5

Assess tree coverage and importance using the line intercept method.

1. Familiarize your team with the species list of trees in table 10.1.
2. Refer to Procedure 10.1 and determine the boundaries of a suitable area to sample trees. Trees include single-stemmed plants with DBH > 2.5 cm.
3. Consult with your instructor and choose a suitable transect length (30–100 m recommended) and a suitable number of transects (one per team recommended).
4. For your transect, obtain a measured 50-m string or rope marked in 10-m intervals.
5. Your transect must be randomly placed. Follow the procedure in step 4 of Procedure 10.3 to randomly place your transect on an imaginary grid with numbered positions encompassing the area being sampled.
6. Stretch the 50-m measured rope on the ground from your random set of coordinates.
7. Along the first 10-m interval, count trees that touch, overlie, or underlie the line. For each tree, record in table 10.9 the species and the length (m) of the line that the tree intercepts. Similarly, for trees that overhang the line, record in table 10.9 the length of the imaginary vertical plane from the line that the tree intercepts. Also record as *Species Bare* any uncovered (bare) lengths of the interval. Table 10.9 accommodates data for seven individuals of each of three species.
8. When all trees from all five 10-m intervals have been recorded, summarize your raw data in table 10.10.
9. Sum the data in both columns and record the totals in table 10.10.
10. Check that each of the other teams collected the data for its transect.
11. Use the data in table 10.10 to calculate the following four parameters for each tree species within the community. Record your results in table 10.11.

    $\text{frequency}_{\text{species}_i}$ = (100 × number of sampling intervals with one or more individuals of species_i) / total number of sampling intervals

    $\text{relative density}_{\text{species}_i}$ = (100 × total number of individuals of species_i) / total number of individuals of all species

    $\text{relative coverage}_{\text{species}_i}$ = (100 × total intercept length by species_i) / total length intercepted by all plants

    $\text{importance value}_{\text{species}_i}$ = ($\text{relative density}_{\text{species}_i}$ + $\text{relative coverage}_{\text{species}_i}$) / 2

12. Your instructor may ask you to compare or combine your data with those from other transects.

Table 10.9

Raw data for trees intercepting a transect at each interval

Extend this table if necessary to accommodate more species.

	Interval 1		Interval 2		Interval 3		Interval 4		Interval 5	
	Individual	Intercept (m)	Individual	Intercept (m)	Individual	Intercept (m)	Individual	Intercept (m)	Individual	Intercept (m)
Species:	1st		1st		1st		1st		1st	
	2nd		2nd		2nd		2nd		2nd	
	3rd		3rd		3rd		3rd		3rd	
	4th		4th		4th		4th		4th	
	5th		5th		5th		5th		5th	
	6th		6th		6th		6th		6th	
	7th		7th		7th		7th		7th	
Species:	1st		1st		1st		1st		1st	
	2nd		2nd		2nd		2nd		2nd	
	3rd		3rd		3rd		3rd		3rd	
	4th		4th		4th		4th		4th	
	5th		5th		5th		5th		5th	
	6th		6th		6th		6th		6th	
	7th		7th		7th		7th		7th	
Species:	1st		1st		1st		1st		1st	
	2nd		2nd		2nd		2nd		2nd	
	3rd		3rd		3rd		3rd		3rd	
	4th		4th		4th		4th		4th	
	5th		5th		5th		5th		5th	
	6th		6th		6th		6th		6th	
	7th		7th		7th		7th		7th	
Species: bare	1st		1st		1st		1st		1st	
	2nd		2nd		2nd		2nd		2nd	
	3rd		3rd		3rd		3rd		3rd	
	4th		4th		4th		4th		4th	
	5th		5th		5th		5th		5th	
	6th		6th		6th		6th		6th	
	7th		7th		7th		7th		7th	

Table 10.10

Summary of raw data for tree species intercepting a transect

Transect length = ________ Number of intervals = ________

$Species_i$	Total Number of Individuals $Species_i$	Total Transect Length Intercepted by $Species_i$
	Total number individuals of all species =	Total intercept length by all plants =

Table 10.11

Relative values of each tree species in a selected community using parameters of the line intercept method

Species	Frequency (%)	Relative Density (%)	Relative Coverage (%)	Importance Value

Questions 7

Which tree species were most evenly distributed among all transect intervals? ______________________

____

Did your sampling of trees reveal any tree species that were not recorded during Procedure 10.1? ______________________

____

Which plant group (herbs, shrubs, trees) was best sampled by quadrats? ______________________

____

Is "best sampled" and "most easily sampled" the same thing? Why or why not? ______________________

____

STRATIFED RANDOM SAMPLING ALONG A GRADIENT

Assessing a plant community, such as herbs in an open field, is often complicated by gradients of resources or habitat features. For example, moisture, soil type, distance from a river, elevation, or distance from a disturbance often vary from one end of a study site to the next. These gradients affect plants and their distributions. The previous procedures have dealt with this variation by randomizing the placement of sampling units throughout the environment. Another sampling procedure that recognizes the realities of the environment and maximizes the information from the samples is **stratified random sampling** (fig. 10.5).

In a stratified random sampling design, the habitat being studied is stratified into meaningful units along a

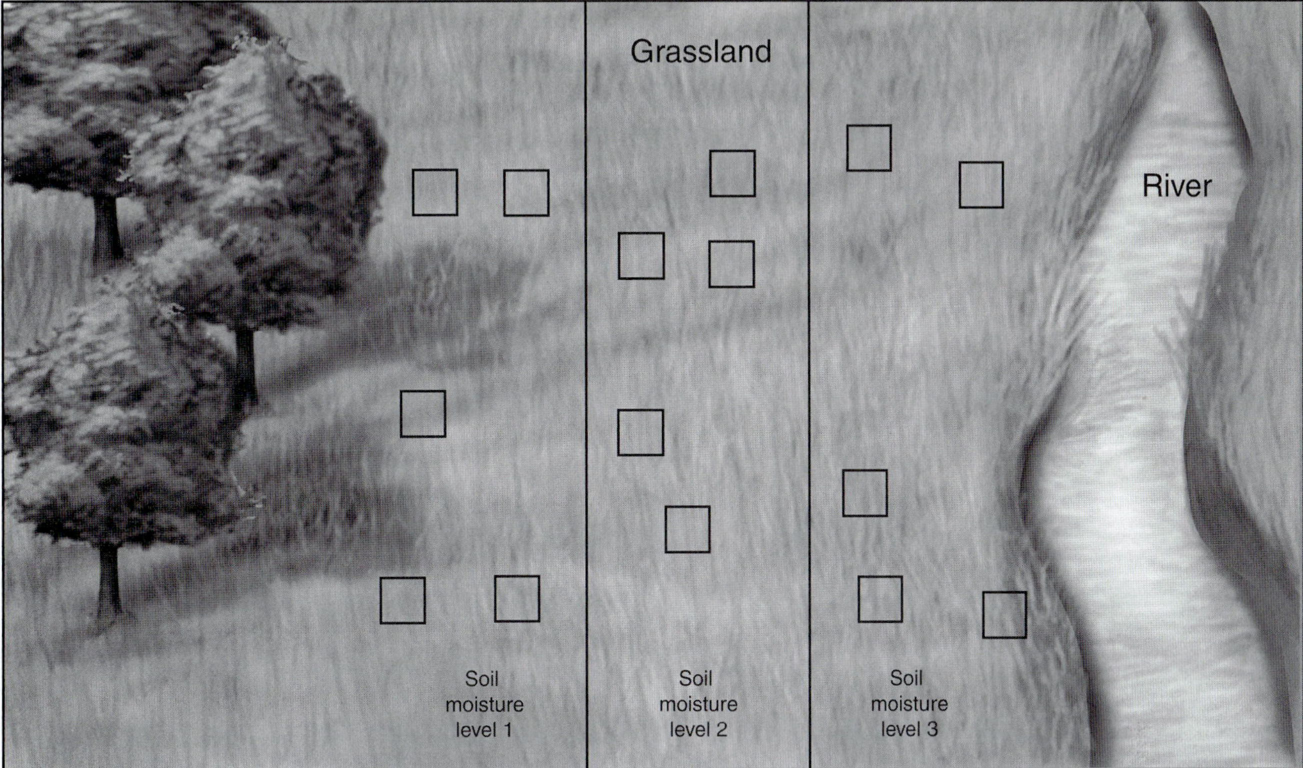

Figure 10.5

For random stratified sampling, a habitat is stratified (divided) into subhabitats along a known environmental gradient such as moisture from an adjacent stream. Samples are randomly distributed within each subhabitat. Stratified sampling accounts for variation of a known environmental gradient.

known gradient. For example, a grassland bordered on one side by a river may have a moisture gradient. If so, the grassland can be stratified into two or more contiguous areas (strata), each a different distance from the stream. Each of the areas is sampled independently and the strata are later compared to determine if the gradient has an effect. Each of the three strata (sub-environments) along the moisture gradient is sampled with its own randomly placed sampling units. Each stratum (defined area along the gradient) is sampled as described in the previous procedures, and the results provide accurate information for areas along the gradient.

Procedure 10.6

Compare plant communities stratified along a gradient.

1. Review the observations of the team that surveyed boundaries and gradients of abiotic factors in Procedure 10.1. Consult with your instructors and review the results of the previous four procedures.
2. Identify a gradient suitable for analysis.
3. Determine which plant type (herb, shrub, tree) to survey.
4. Choose the most appropriate sampling design (quadrat, line transect, strip transect) and randomization procedure.
5. Divide the habitat into biologically meaningful strata along the gradient.
6. Design the tables needed for recording raw data, summarizing raw data, and calculation of the variables needed to describe the community.
7. Divide into teams (one team for each stratum) and sample the designated plant community according to the steps in the previous procedures.
8. Calculate and compare your results for the strata.

Questions 8

After completing your sampling, do your original divisions of the habitat into strata still appear biologically meaningful? ______________________________

____

What did you conclude about the effect of the gradient on herbs? ______________________________

____

On shrubs? ______________________________

____

On trees? ______________________________

____

Questions for Further Thought and Study

1. Evaluate the advantages and disadvantages of

   quadrat sampling:

   strip transect sampling:

   line transect sampling:

2. Would a greater number of quadrats used in a sampling design narrow or widen the confidence intervals around a mean density? Why? How might you investigate this by subdividing your data set from Procedure 10.2?

3. How would you design an experiment to test the relative effectiveness of circular versus square versus rectangular quadrats?

4. The importance of a plant species to a community is difficult to define. What would be five or more meaningful parameters to measure?

11 Sampling Animal Communities

Objectives

As you complete this lab exercise you will:

1. Estimate the size of a simulated population using the Lincoln-Petersen calculations of mark-recapture data.
2. Use mark-recapture to determine insect population size.
3. Use variable circular plots to determine the density of a bird population.
4. Trap small mammals with Sherman live traps to determine the number of species and their relative abundance in a small mammal community.
5. Use animal sampling data to calculate Jacquard's coefficient of community similarity and investigate minimum and maximum similarity values.
6. Choose and conduct a data-collection technique (sweep net, mammal trapping, etc.) to determine similarity between two communities.

Figure 11.1
A sweep net is a standard collecting device that is ruggedly built with canvas and netting. Sampling effort with a sweep net is quantified with a consistent number of sweeps at a consistent speed.

Populations of animals are hard to sample because they move (fig. 11.1). However, the same principles of good design used in plant sampling also apply to animal sampling. Careful planning, familiarity with the area, and careful observation are crucial. The best procedures include taking enough repeated samples to estimate population values accurately. Most procedures to sample animal populations gather data to answer questions about population size, density, diversity, similarities among communities, and changes in these parameters through time.

This lab exercise presents three common procedures for sampling populations of invertebrates and small vertebrates: mark recapture, circular plots, and mammal live trapping. Data analyses for these procedures focus on calculating population size and community similarity. However, data from these techniques also allow calculation of a variety of parameters and indices. The best sampling procedures provide enough information to broaden your research.

MEASURING POPULATION DENSITY WITH MARK-RECAPTURE

Sampling population size of highly mobile species is difficult at best. Fortunately, an ingenious procedure to overcome the movement problem has been around for over 200 years—a procedure called **mark-recapture**. For this procedure, a sample of animals is captured by whatever means. All of the individuals are marked with a number, a notch, or a tag, and then released. A second sample is taken some days later. Some of the individuals captured in the second sample were marked from the first sample. The size of the two samples and the portion of the second sample that is marked allow calculation of total population size.

The most common calculations for mark-recapture follow the **Lincoln-Petersen method**:

$$P = (M \cdot p)/m$$

where:

P = total population size

M = total number of marked individuals (= size of the first sample)

p = size of the second sample

m = number of marked individuals in the second sample.

Procedure 11.1

Estimate population size using the Lincoln-Petersen calculations of mark-recapture data.

1. Examine the following simulated mark-recapture data. These data include counts of fish that were marked, released, and recaptured.

   The first sample of fish contained 250 individuals; all were marked and released.

   The second sample of fish contained 300 individuals, 25 of which were marked.

2. Calculate the total population size:

   $P = (M \cdot p)/m$

   P = total population size is _______ individuals.

3. Verify your calculations with your instructor.

Question 1

If you used mark-recapture to estimate population size for a grasshopper population and for a fish population, could you relate population size to population density for either of these populations? How so? ________________________

____

Mark-recapture sampling applies to a variety of species and habitats. Like all procedures, however, certain assumptions must be met for valid results. Mark-recapture assumes that:

- Animals retain their marks.
- Marks do not alter natural behavior or survival rates.
- Marked animals have the same probability of being recaptured as do unmarked animals.
- Marked animals distribute themselves randomly among unmarked animals.
- Emigration and immigration are equal for marked and unmarked individuals.

Procedure 11.2

Use the mark-recapture procedure to determine population size of grasshoppers.

1. Consult with your instructor to locate a tall-grass habitat with an abundant grasshopper population (fig. 11.2).

Figure 11.2

This grasshopper is being marked for recapture. Mark-recapture methods determine population size, but are also part of behavioral studies.

2. Walk the habitat and take note of significant variations that might influence the distribution and abundance of grasshoppers and the placement of your sampling sites.
3. Use an insect sweep net to collect a sample of grasshoppers from the area. After examining the sample, consult with your instructor about marking all species of grasshopper, or a single abundant species. Determine the smallest grasshopper (≈ 1 cm) that can be marked on top of the thorax with a blue or red dot from a felt-tip marker.
4. Designate sampling sites about 10–20 m apart throughout the habitat. Record the sampling site IDs in table 11.1. Establish teams of two to four students and designate one sampling site for each team.
5. Obtain an insect sweep net to sample grasshoppers at your team's site.
6. To collect grasshoppers, "sweep" the net back and forth so the hoop brushes through the grass rather than above it. Sweep the net five to seven times. Pass each sweep through a new patch of grass. Be consistent with speed and movement. If populations are low, more sweeps might be necessary.
7. Concentrate the captured insects into the bottom of the net by shaking it, but don't let them escape.
8. Grab and constrict the net just above the concentrated insects to prevent them from escaping and invert the end of the net into a wide-mouth jar. Replace the lid. This takes practice. Watch your instructor do this.

9. Examine the captured insects for grasshoppers suitable for marking.
10. Patiently take each grasshopper out of the jar, mark it on top of the thorax, and put it in a holding container. On a raw data sheet, place one tally mark for each marked and released grasshopper.
11. Repeat steps 6–10 until you have marked 50–100 grasshoppers.
12. Release all of the marked grasshoppers.
13. Sum the tally marks and record for your sample site in table 11.1 your *Number of Marked Grasshoppers* for the first sample. Also record the totals from the other teams' sampling sites.
14. Leave the sampled habitat undisturbed for 1 or 2 days. Then return for a second sample.
15. Repeat steps 4–8 until you have captured 50–100 grasshoppers. Accumulate your second sample in a holding container.
16. Count and record in table 11.1 the *Total Number of Grasshoppers Captured* in your second sample.
17. Count and record in table 11.1 the *Total Number of Marked Grasshoppers* in the second sample.
18. Release the second sample of grasshoppers.
19. Record in table 11.1 the totals for the second samples by the other teams.
20. Use the Lincoln-Petersen method (Procedure 11.1) to calculate the *Total Population Size Estimate* and record it in table 11.1.

Questions 2

Was the grasshopper population smaller than you expected? ______________________________

____

Are you confident that your procedure to measure grasshopper population size met all the assumptions for successful mark-recapture? ______________________________

____

What assumptions are most suspect for your procedure?

____

____

Does sweep netting bias your samples? How so? __________

____

Should marks be as easy to see as possible to make sure you don't miss one? Why? ______________________________

____

What weaknesses in mark-recapture would you try to minimize? ______________________________

____

VARIABLE CIRCULAR PLOTS TO SAMPLE BIRD POPULATIONS

The **variable circular plot** method works well for censusing animals by sight and sound. In this method, a stationary observer records reliable sightings of animals appearing in any direction from a central point. This makes the plot circular (fig. 11.3). The radius of the plot ultimately depends on the farthest reliable sighting, which varies with density of the vegetation. Distances to sighted animals are recorded and may be any length. Determining the distance of reliable detection (i.e., the radius of the circle and the area of the plot) requires a graph of the raw data showing the distance at which detection drops off.

Procedure 11.3

Circular plot procedure to determine the density of a bird population.

1. Choose with your instructor a consistent, typically wooded habitat with an abundant bird population.

TABLE 11.1

MARK-RECAPTURE DATA FOR DETERMINING THE SIZE OF A GRASSHOPPER POPULATION

Sampling Site ID	Number of Marked Grasshoppers (first sample)	Total Number of Grasshoppers Captured (second sample)	Number of Marked Grasshoppers (second sample)
	Total:	Total:	Total:
		Total population size estimate:	

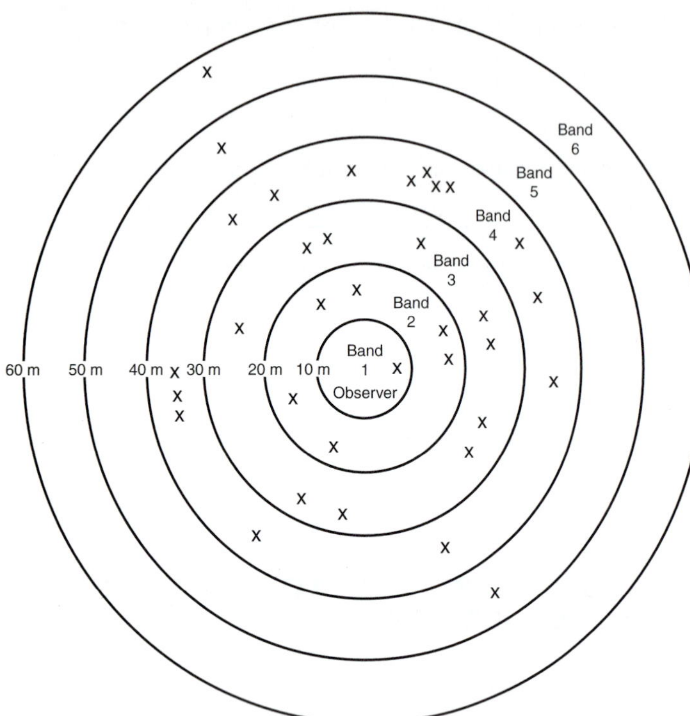

Figure 11.3

Variable circular plot. The plot for this sample site includes six bands. Each X marks an observation of an organism. The number of observations is sharply lower in the two outermost bands. Therefore, they are not considered reliable detection bands.

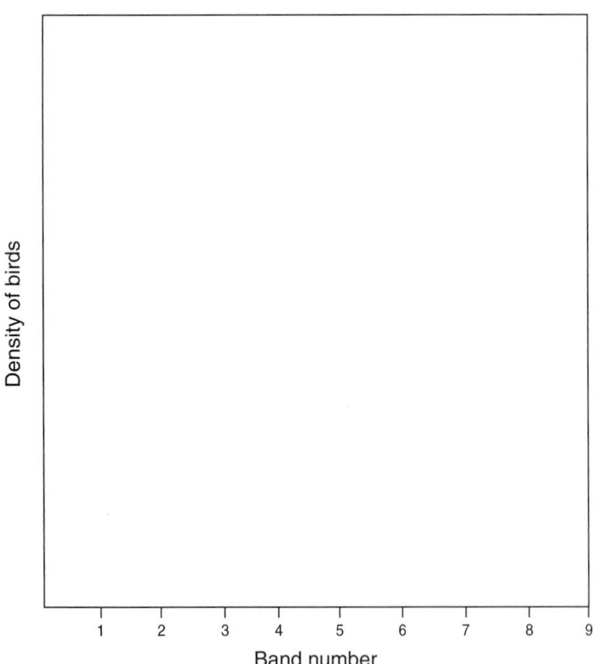

Figure 11.4

Graph for density of birds versus band number for student-collected data.

2. Establish sampling stations (one for each student or pair of students) along a transect covering the entire habitat. Place the sampling stations 50–200 m apart depending on the habitat available and the potential sight-line from each observer. The distance should prevent two observers from seeing the same bird during the same time interval.
3. At one or all sampling stations, mark 10-m distance bands (intervals) to help the observer learn to gage distances.
4. Position yourself as a stationary observer at the center of a 360° circular sampling station. As you walk to the center, count and record the distance from the station center of any bird seen flying away.
5. As an observer, sample for 15 min and record in a field notebook each bird you have seen or heard and its estimated distance away.
6. There is no maximum distance. Record all birds except those flying far overhead that are not using the general habitat.
7. Transcribe your field notebook data into table 11.2.
8. Record in table 11.2 the other teams' data from their sampling stations.
9. For each band, calculate and record the *Total number of birds* by summing the bird counts from all stations.
10. To determine the area sampled reliably, first calculate for each band the *Mean number of birds per station* by dividing the total number of birds by the number of stations. Record these values in table 11.2.
11. Calculate and record in table 11.2 the *Density of birds for each band* as:

    density of birds in band_i = (mean birds per station for band_i) / (area m^2 of band_i)

12. To determine the visual area that is most probably sampled accurately, plot in figure 11.4 the *Density of birds* for each band versus *Band number*.
13. Examine your graph in figure 11.4 and determine the band at which bird density drops off sharply. This is the inflection point. The concentric bands before the inflection point represent the area with reliable bird counts. They are the *reliable detection bands*. Indicate the *Reliable detection bands* with a *Yes* or *No* in table 11.2.
14. Determine the area of the circular plot with a radius that includes all reliable detection bands by summing the *Area of band* values for all columns marked as *Reliable detection bands*. Record this value as *Total area (km^2) of reliable detection bands*.
15. Sum for all sampling stations the number of birds sighted or heard within each reliable detection band. Divide this value by the number of stations and record it as *Mean number of birds within all reliable detection bands per station*.

TABLE 11.2

DATA AND CALCULATIONS FOR USING THE VARIABLE CIRCULAR PLOT METHOD TO DETERMINE A BIRD POPULATION DENSITY

	Number of birds seen or heard in 15 minutes								
Sampling Station	Band 1 0–10 m	Band 2 > 10–20 m	Band 3 > 20–30 m	Band 4 > 30–40 m	Band 5 > 40–50 m	Band 6 > 50–60 m	Band 7 > 60–70 m	Band 8 > 70–80 m	Band 9 > 80–90 m
Total number of birds									
Mean number of birds per station									
Density of birds for each band									
Reliable detection band (Y/N)									
Area of band	314 m^2	942 m^2	1571 m^2	2199 m^2	2827 m^2	3456 m^2	4084 m^2	4713 m^2	5340 m^2
Total area (km^2) of reliable detection bands = ______	Mean number of birds within all reliable detection bands per station = ______					Total bird population density = ______ km^{-2}			

16. Calculate and record in table 11.2 the *Total bird population density* by dividing the *Mean number of birds within all reliable detection bands per station* by the *Total area of reliable detection bands.*

Questions 3

Why shouldn't the observation time be as long as possible to see as many birds as possible? ____________________

____

What weaknesses in variable circular plot procedures would you try to minimize? ____________________

____

SAMPLING A POPULATION OF SMALL MAMMALS

Sherman live traps distributed in the field can effectively sample a population of small mammals. These traps are typically baited and distributed over a grid large enough to cover the study area. Traps are typically set to capture animals at night when they are most active (fig. 11.5).

Procedure 11.4

Use Sherman live trapping to determine the number of species and their relative abundances in a small-mammal community.

1. Discuss with your instructor the ethical treatment of animals and the safety risks involved in small-mammal trapping.
2. Watch your instructor operate a Sherman live trap.
3. Consult with your instructor and locate a community for sampling with a grid of Sherman traps.
4. Use a 100-m tape to lay out a grid to cover the sampling site. The grid does not have to be a square. It should cover all contiguous areas with consistent habitat. The lines of the grid should intersect at 10-m intervals.
5. Sketch the grid and number the intersections. A 50-m × 50-m grid has 36 intersections.
6. The traps should be set during late afternoon and checked the next morning. A grid of 36 traps set and checked after each of two nights constitutes 72 trap nights.
7. Place one trap at each intersection. For each trap:
   a. Bait the trap with dry oats or with a small amount of peanut butter mixed with dry oats. Do not use peanut butter if experience has shown that ants are a problem in the sampling area. Ants can kill a trapped animal.
   b. Set the trap with the door open and in the shade of leaves or a rock if possible. Covering the top of the trap with leaves will help insulate the trap. Your instructor may advise you to include cotton nestlets to help insulate a small mammal on a cold night.
   c. If a grid intersection lies on an incline, face the trap door uphill.
8. Let the traps remain undisturbed overnight.

Figure 11.5

A Sherman trap is easily set by pushing the spring-loaded door down to hook against a latch plate. An animal enters and must walk across the plate to get the food inside. Its weight releases the plate and the door swings up for live capture.

9. In the morning, check the traps for sprung doors. If the door is still open, close it. Don't leave traps open during the day.
10. If a trap door is sprung closed, gently detect if a small animal is inside.
11. If an animal is in the trap, open a pre-weighed dark cloth bag, put the mouth of the bag over the sprung door, and open the door facing up to coax the animal to crawl out.
12. Close the bag, identify the animal, and record the species in table 11.3. Also record in column *Number of Individuals* one tally mark for each trapped individual of that species. Consult with your instructor about gathering weight and gender data.
13. Gently release the animal, return the trap to its grid position, and close the door.
14. Be sure to check ALL the traps. Never let an animal remained trapped for more than 12–15 h.
15. After checking all the traps and closing the doors, leave them in position until you return that afternoon to begin another trapping night.
16. For the next trapping night, return to the grid in the afternoon. Follow steps 7–15.
17. Each morning for three days, gather and record your data in table 11.3.
18. Calculate and record in table 11.3 the *Cumulative number of species for all dates* and the *Number of species trapped* for each date.
19. For each species, sum the number caught for all nights and record that number in the column for *Total abundance of species*$_i$.
20. Record in the *Species Rank* column of table 11.3 the names of the species in order from most abundant to least abundant. Record next to the ranked species names *Total Abundance of Each Ranked Species*. These are simply reordered values from a previous column.
21. Your instructor may ask you to graph the abundances of each species as a rank abundance curve (see Procedure 12.3).

Questions 4

Are the species equally abundant? How many species comprise 90% of the individuals? ______________________

____

Table 11.3

Data from trapping small mammals to determine the number of species in a small-mammal population

	Date:	Date:	Date:			
Species List	Number of Individuals	Number of Individuals	Number of Individuals	Total Abundance of Species$_i$ for All Dates	Species Rank	Total Abundance of Each Ranked Species
Cumulative number of species for all dates = ______	Number of species trapped = ______	Number of species trapped = ______	Number of species trapped = ______			

Small mammals sometimes become "trap happy" and are caught over and over again. Why would this be so common? ______________________________

____

How would you minimize the bias that trap-happy mammals introduce to your data? ______________________________

____

What factors limit this procedure's sensitivity to all mammals present in the community? ______________________________

____

What weaknesses in small-mammal trapping procedures would you try to minimize? ______________________________

____

Could small-mammal trapping be part of a mark-recapture procedure? What would be the greatest concerns? ________

____

JACQUARD'S COEFFICIENT OF COMMUNITY SIMILARITY

One powerful way to learn about a community is to compare its species composition to that of another community. Ecologists compare diversity, tree dominance, food chains, species lists, and a myriad of other parameters to find patterns that tell us how a community is put together and how it "ticks." One group of metrics, called **community similarity indices**, compares communities based on the number of species in each community and the number of shared species. A common measure of similarity is **Jacquard's coefficient of community similarity**, defined as:

$$CS_J = c / (s_1 + s_2 + c)$$

where:

CS_J = Jacquard's community similarity coefficient

c = the number of species common to both communities

s_1 = number of species in community 1 but not in community 2

s_2 = number of species in community 2 but not in community 1

Jacquard's coefficient uses presence/absence data. Abundance data isn't needed.

Procedure 11.5

Calculate Jacquard's coefficient of community similarity and investigate minimum and maximum values.

1. Examine the columns of *Simulated Data for Calculation of* CS_J in table 11.4 for tree species in two communities.
2. Calculate and record in table 11.4 the number of tree species unique to each community (s_1 and s_2), the number of species common to both communities (c), and Jacquard's coefficient (CS_J).
3. Verify your calculations with your instructor.
4. Propose and record in table 11.4 two simulation data sets—one that will result in the minimum possible coefficient of similarity, and one that will result in the maximum coefficient.
5. Calculate CS_J for both proposed data sets and record them in table 11.4.

TABLE 11.4

SIMULATED DATA FOR PRACTICE CALCULATIONS OF CS_J AND TWO BLANK FORMS FOR SIMULATION OF COMMUNITIES WITH MINIMUM SIMILARITIES AND WITH MAXIMUM SIMILARITIES

Simulated Data for Calculation of CS_J		Proposed Data to Calculate Minimum Possible Similarity		Proposed Data to Calculate Maximum Possible Similarity	
Tree species in community 1	Tree species in community 2	Community 1	Community 2	Community 1	Community 2
Species A	Species G				
Species B	Species B				
Species C	Species C				
Species D	Species F				
Species E					
s_1 =	s_2 =	s_1 =	s_2 =	s_1 =	s_2 =
Number of species in common = c =		Number of species in common = c =		Number of species in common = c =	
Similarity coefficient CS_J =		Minimum similarity coefficient CS_J =		Maximum similarity coefficient CS_J =	

Questions 5

What is the maximum possible value for Jacquard's coefficient of similarity? Minimum?________________________

____

Does the use of a single taxon such as trees, birds, or fish rather than *all* species bias your conclusions about the similarity of two communities? If so, how? ______________

____

What taxon would you suggest is the most indicative of a community's structure? Why? ________________

____

Procedure 11.6

Choose and conduct a data-collection technique (sweep net, mammal trapping, etc.) to determine similarity between two communities.

1. Consult with your instructor and choose two communities to compare.
2. Choose a taxon as your basis for comparison of the communities.
3. Choose a sampling technique and design your experiment.
4. Record your data in a field notebook and transcribe the information into table 11.5.
5. Calculate the number of species in each community, the number of species common to both communities, and Jacquard's coefficient of community similarity.

TABLE 11.5

DATA FOR CALCULATION OF SIMILARITY BETWEEN TWO COMMUNITIES

Raw Data for Calculation of CS_J	
Species in Community 1	Species in Community 2
Number of species = s_1 =	Number of species = s_2 =
Species in common, c =	
Jacquard's coefficient of community similarity = CS_J =	

Questions for Further Thought and Study

1. Each technique to sample animal populations and communities has inherent error and bias. How would you minimize that error in most cases?

2. Which of the techniques presented in this lab exercise offers the most information about the population? Is the amount of information a function of the procedure?

Species Diversity

12

Objectives

As you complete this lab exercise you will:

1. Calculate species richness, Shannon-Wiener diversity, H_{max}, evenness, and Simpson's diversity index.
2. Calculate and graph a rank-abundance curve.
3. Compare insect diversity in two contrasting terrestrial environments.
4. Compare fish diversity in two contrasting aquatic environments.

Species diversity is among the most calculated and cited of all ecological variables. No other measure provides more insight into complex communities. Diversity reflects stability, aesthetics, and health of ecosystems (fig. 12.1). Ecologists define and calculate **species diversity** based on two characteristics: (1) the number of species in the community, which ecologists call **richness**; and (2) the relative abundance of species, called **evenness**.

In this lab exercise you will calculate a variety of diversity indices and graph relative abundance to visualize the diversity of organisms in a community or ecosystem. You'll apply diversity calculations to samples of insects collected with a sweep net and to fish collected with a seine net.

SPECIES RICHNESS

Richness is the easiest measure of diversity because it's simply a count of the species present. Everyone agrees that the most diverse communities have the most species. But, using richness to measure diversity assumes that one can record *all* species in a community, even the rare ones. That's difficult to do. Therefore, richness can be misleadingly correlated with sample size—the greater the sample size, the more likely rare species are found. Nevertheless, recording richness is a good first step to measure community diversity.

$$\text{richness} = \text{number of species} = s$$

Figure 12.1

Tropical forests are among the most diverse ecosystems in the world. This upland forest in the Galápagos Islands lies on the equator and includes an unusual variety of ferns, trees, herbs, and epiphytes.

SPECIES EVENNESS

Richness clearly relates to community diversity—a community with 80 species is obviously more diverse than one with 20 species. But evenness is also meaningful. For example, the data in table 12.1 contrast two forest communities with equal richness. Examine the *Species* and *Counts* columns. Both contain five tree species. However, community *b* is more diverse than community *a* because the relative abundance of species

TABLE 12.1

EXAMPLE DATA FOR RICHNESS, EVENNESS, AND DIVERSITY

Community *a*		Community *b*	
Species	Count	Species	Count
1	21	1	5
2	1	2	5
3	1	3	5
4	1	4	5
5	1	5	5
Total abundance = 25		Total abundance = 25	

is more even for community *b*. In community *b*, all five species are equally abundant, each comprising 20% of the tree community. In contrast, 84% of the trees in community *a* belong to one species, whereas each of the remaining species is only 4% of the community. On a walk through the two forests, you would certainly sense more diversity in community *b*, despite equal species richness in the two forests. For this reason, most indices of diversity combine information of evenness (relative abundance) with richness.

Questions 1

In your experience, which is more common—a community dominated by a few species or a community with all species represented equally? ______________________________

____

Species richness may sometimes correlate with sampling effort. Why is this a problem for accurately assessing a community? ______________________________

____

SHANNON-WIENER INDEX OF SPECIES DIVERSITY

Ecologists have developed many species-diversity indices with components of species richness and of evenness. One of the most commonly used indices is the Shannon-Wiener index, which states: the higher the index value, the greater the diversity.

$$H' = -\sum_{i=1}^{s} p_i \ln p_i$$

where:

H' = Shannon-Wiener diversity index

p_i = the proportion of the ith species = abundance of species$_i$ / total abundance of all species

$\ln p_i$ = the natural logarithm of p_i

s = the number of species in the community

To calculate H', determine the proportions (p_i) of each species in the community, then the ln of each p_i. Then multiply each p_i times $\ln p_i$ and sum the results for all species from species 1 to species s, where s = the number of species in the community. This sum is negative, so take the absolute value to complete the calculations. Examine table 12.2 for example calculations.

The minimum value of H' is 0, which is the value for a community with a single species. Values range to ∞, but 7 denotes an extremely rich community. Communities with a Shannon-Wiener diversity of 1.7 or higher are considered relatively diverse. The maximum value increases as species richness and species evenness increase. H'_{max} is defined as:

$$H'_{max} = \ln s$$

AN INDEX OF SPECIES EVENNESS

An index of evenness (J') can be derived from the Shannon-Wiener index. This index of evenness ranges from 0–1, and the value is most meaningful for comparisons between communities rather than as a stand-alone measure. An index of evenness reveals the extent that a community is dominated by only a few of its species (uneven) or has all of its species in similar numbers (even). Evenness can be defined as:

$$J' = H' / H'_{max}$$

where:

H' = Shannon-Wiener diversity index

$H'_{max} = \ln s$

s = the number of species in the community

SIMPSON'S INDEX OF SPECIES DIVERSITY

Simpson's index is another common measure of diversity. As in the Shannon-Wiener index, the calculations are based on the proportion of each species in the total sample. Simpson's index weighs dominant species somewhat more than rare species in its measure of diversity than does the Shannon-Wiener index. Simpson's index tends to vary less between samples, and it ranges from 0 to ∞. Values are commonly a few points higher than calculations of the Shannon-Wiener index. Simpson's index is defined as:

$$D = 1 / \sum(p_i^2)$$

where:

D = Simpson's index

p_i = the proportion of the ith species

Table 12.2

Example data for richness, evenness, and Shannon-Wiener diversity index (H') calculations for two hypothetical forest communities

Different values of H' for the two communities reflect different species evenness. H' for community *b*, the community with higher species evenness, is 1.610; H' for community *a* is 0.662.

Community *a*					Community *b*				
Species	Count	Proportion (p_i)	$\ln p_i$	$p_i \ln p_i$	Species	Count	Proportion (p_i)	$\ln p_i$	$p_i \ln p_i$
1	21	0.84	−0.174	−0.146	1	5	0.20	−1.609	−0.322
2	1	.04	−3.219	−0.129	2	5	0.20	−1.609	−0.322
3	1	.04	−3.219	−0.129	3	5	0.20	−1.609	−0.322
4	1	.04	−3.219	−0.129	4	5	0.20	−1.609	−0.322
5	1	.04	−3.219	−0.129	5	5	0.20	−1.609	−0.322
Total abundance	25	1.00	H' =	0.662	Total abundance	25	1.00	H' =	1.610

Question 2

Could a community of 15 species have the same diversity as measured by Shannon-Wiener or Simpson's index as a community with 30 species? Why or why not? ____________

____

____

RANK-ABUNDANCE CURVES

A **rank-abundance curve** (fig. 12.2) makes information about diversity, richness, and evenness accessible at a glance. In this graph, the rank of each species is plotted along the x axis. The most abundant species is ranked 1, the second most abundant species is ranked 2, and so forth. The abundance of each species is plotted on the log scale of the y axis. The shape of the curve reveals subtleties about community structure. High, flat curves indicate high diversity. Low and steep curves reveal low diversity and greater dominance by only a few species. Environmental stress usually steepens the curve for a community.

Questions 3

Why would stress steepen a relative-abundance curve?

____

____

Rank-abundance curves are constructed with data from sampling a community. How would more samples likely change the curve? ____________

____

Procedure 12.1

Calculate species richness, Shannon-Wiener diversity index, H'_{max}, and evenness from simulation data.

1. Examine the example data and calculations in table 12.2.
2. Examine the raw data in table 12.3.

Table 12.3

Raw data set for calculation of richness, Shannon-Wiener diversity, and species evenness

Species	Abundance	Rank	Proportion (p_i)	$\ln p_i$	$p_i \ln p_i$
A	40				
B	28				
C	254				
D	14				
E	225				
F	150				
G	75				
H	110				
I	101				
J	95				
Richness =	Shannon-Wiener index =		H'_{max} =		Evenness =

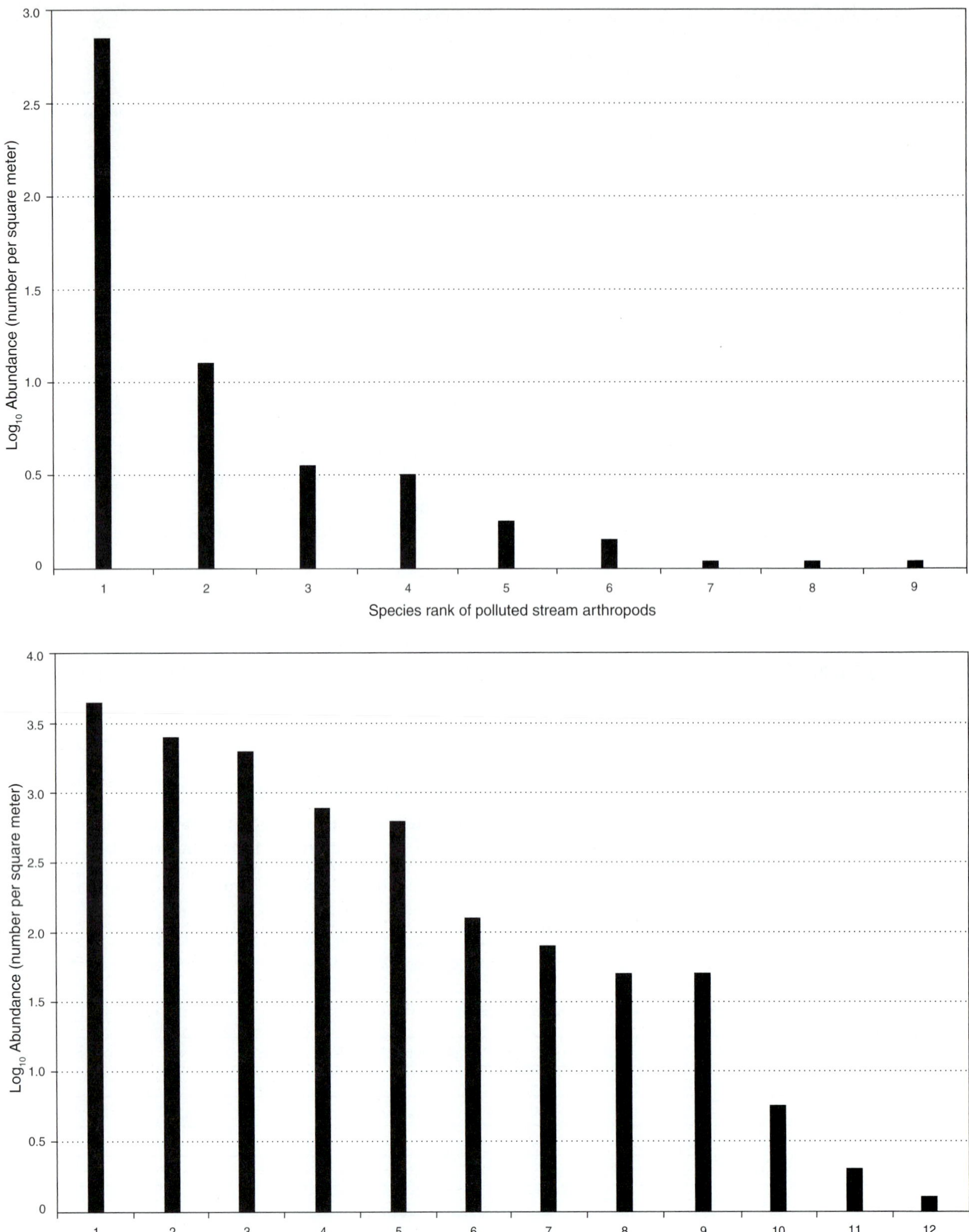

Figure 12.2

Two contrasting rank-abundance curves for (a) benthic (sediment dwelling) arthropods in a polluted, silt-bottom stream; (b) benthic arthropods in a rocky-bottom, mountain stream.

3. Calculate richness, Shannon-Wiener diversity, H'_{max}, and species evenness for the data in table 12.3.
4. Record your results in table 12.3 and verify your results with your instructor.

Procedure 12.2

Calculate Simpson from a raw data set.

1. Examine the raw data table 12.4.
2. Indicate the rank of each species in the *Rank* column. The most abundant species is ranked 1, etc.
3. Calculate and record in table 12.4 the proportion (p_i) of the total for each species, then calculate the squared proportions of each species (p_i^2).
4. Sum the squared proportions.
5. Calculate and record Simpson's diversity for these data. Verify your calculations with your instructor.

Question 4

Simpson's index is 1.4 for community *a* in table 12.1. What is it for community *b*? ______________________

____

Procedure 12.3

Calculate a rank-abundance curve from practice data.

1. Examine the raw data in table 12.4 and the ranking of each species.
2. Plot in figure 12.3 the $\log_{10}$ *of the abundances* versus rank for each species.

Questions 5

Do any steep parts of the curve indicate dominance by just a few species? ______________________

____

Some communities have many rare species. How would the curve be shaped for such a community? ______________

____

INSECT DIVERSITY AS MEASURED FROM SWEEP NET SAMPLES

Measuring total diversity of a community would, in theory, require counting all species present. This is a monumental

Table 12.4

Data for Calculating Simpson's Diversity Index and for Ranks Used in Producing a Rank-Abundance Curve

Species	Abundance	Rank	Proportion (p_i)	p_i^2
A	50			
B	5			
C	342			
D	7			
E	798			
F	5			
G	503			
H	100			
I	200			
J	327			
K	11			
L	350			
M	1112			
N	339			
O	300			
P	375			
Q	42			
Richness =	Total =			$\Sigma(p_i^2)=$
			Simpson's index = _____	

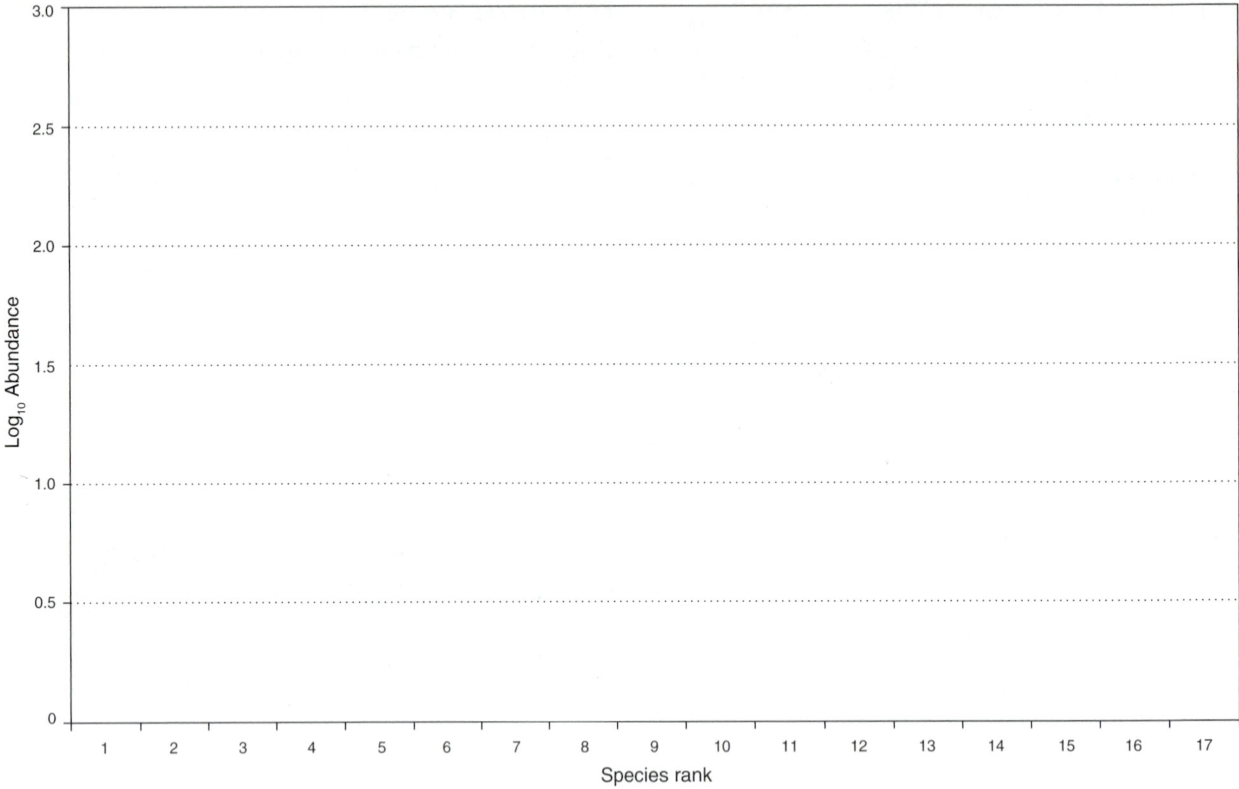

Figure 12.3

Rank-abundance curve for the simulation data in table 12.4.

task, especially if you include the myriad of microorganisms. In practice, ecologists usually sample a narrower taxon—such as trees, plants, vertebrates, or insects—either according to their interests or as a surrogate index of overall community diversity. For a selected taxon to realistically represent the entire community, the taxon should have high inherent diversity. Certainly, the most diverse group of organisms is arthropods, especially insects (fig. 12.4). Therefore, insect species are often used to reflect overall community diversity.

Procedure 12.4

Compare insect diversity in two contrasting environments.

1. Consult with your instructor and choose two contrasting habitats to compare insect diversity. Choose a site in each habitat to sample.
2. Prepare an insect killing jar.
   a. Pack 2 cm of cotton in the bottom of a large (≈ 1 qt) jar with a tight-fitting lid. Pour a 2–3-cm deep layer of plaster of Paris to cover the cotton completely. Let it harden.
   b. Activate the killing jar by pouring 10 ml of ethyl acetate into the jar. It will penetrate the porous plaster of Paris and saturate the cotton.
   c. Replace the lid immediately and tightly. The fumes are lethal to insects.
3. Obtain an insect sweep net.
4. To collect insects, "sweep" the net back and forth so the hoop brushes across the tops of the vegetation. Sweep the net 25 times, with each sweep passing across a new patch of vegetation. Be consistent in speed and movement.
5. The insects accumulated in the net represent one sample from one "unit effort." The unit effort includes 25 sweeps (or the number of sweeps advised by your instructor).
6. Concentrate the caught insects down to the bottom of the net by shaking it, but don't let them escape.
7. Grab and constrict the net just above the concentrated insects to prevent them from escaping.
8. Open the killing jar and invert the end of the net with the trapped insects into the jar and quickly replace the lid. This takes practice. Watch your instructor do this.
9. After 20 min, the insects should be dead. Open the jar and shake them out into another container labeled for site and sample number.
10. Take at least three replicate samples from each site.

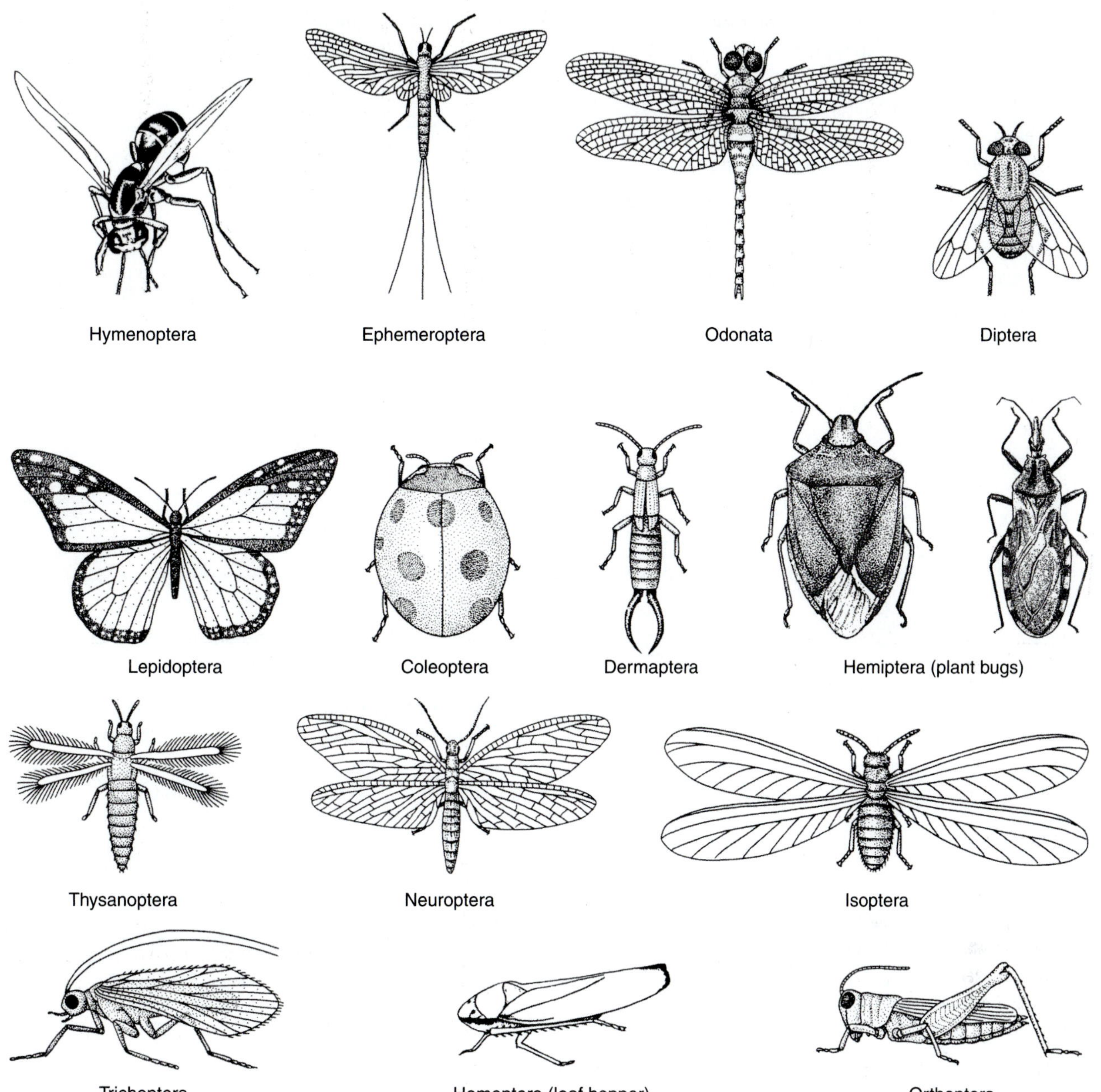

Figure 12.4

The most common insect orders.

11. Before classifying and counting the insects, familiarize yourself with representatives of each Order. Use a practice collection provided by your instructor.
12. For each sample, pour the specimens into a tray and sort them into groups of similar specimens.
13. Examine the specimens with a dissecting microscope, and use the key in the Boxed Reading on page 123 to classify them.
14. Decide with your instructor if you will classify specimens to Order only, or if you will try to group specimens into apparent species based on morphological similarities.
15. Count and record in table 12.5 the insects of each Order (or species).
16. Calculate and record in table 12.5 richness, Shannon-Wiener diversity, H_{max}, evenness, and Simpson's index for each of the two sampling sites.

Questions 6

What was the most abundant order of insects in your samples? ______________________________

____

Were the differences in the two environments reflected in their insect diversities? ______________________________

____

Which sampling site was the most diverse? ______________

____

You chose environments to sample based on some contrasting features. How might those features account for the differences in diversity? ______________________________

____

Did all measures of diversity lead you to the same conclusions? ______________________________

____

Consider the variation among your sweep-net samples. Would more samples likely lead you to different conclusions? Why? ______________________________

____

Ecologists sometimes express the number of organisms as "catch-per-unit-effort" rather than absolute density. What was the mean catch-per-unit-effort for your sweep-net samples for each sampling site? ______________________________

____

Procedure 12.5

Compare fish diversity in two contrasting environments.

1. Consult with your instructor and choose two contrasting aquatic sites to compare fish diversity. They may be a stream pool versus riffle, two streams, or a stream and a lake shore.
2. Obtain a 12-ft (or longer) fish seine net.

Table 12.5

Insect counts, classification, and diversity from sweep net samples of two contrasting sites

Site ______________						Site ______________					
Taxon	Sample 1	Sample 2	Sample 3	Sample 4	Totals	Taxon	Sample 1	Sample 2	Sample 3	Sample 4	Totals
Richness ______	Shannon-Wiener ______	H_{max} ______	Evenness ______	Simpson's ______		Richness ______	Shannon-Wiener ______	H_{max} ______	Evenness ______	Simpson's ______	

A Dichotomous Key to Common Orders of Insects

A common tool to identify organisms is a **dichotomous key.** Dichotomous keys list and describe pairs of opposing traits, each of which leads to another pair of traits until a level of classification of the specimen being identified is reached. By using a key, you'll learn the characteristics that distinguish each of the groups identified by the key.

Insects are classified into 26 Orders distinguished mainly by the structure of wings, mouthparts, and antennae. The key below will help you identify the Order of the insects collected in your sweep net samples. To use the key:

1. Select a specimen and read the first pair of characteristics.
2. Choose the one that best describes your specimen.
3. Proceed according to the number at the end of your choice to the next pair of characteristics.

DICHOTOMOUS KEY TO COMMON ORDERS OF INSECTS

1. Insects with 2 wings .... **(flies) Diptera**
   Insects with 4 wings, a pair of forewings, and a pair of hindwings 2
2. Fore- and hindwings are not alike in texture and color One pair may be hard and dense while the other may be light and transparent 3
   Fore- and hindwings similar, usually clear, thin, and transparent... 5
3. Forewings thick and leatherlike at base, tips much thinner and may be transparent; mouthparts pointed and beaklike to puncture prey and suck body fluids...................................... **(bugs) Hemiptera**
   Forewings same texture throughout, biting mouthparts with opposing mandibles 4
4. Forewings leathery and with veins .... **(grasshoppers, crickets) Orthoptera**
   Forewings hard, without veins.........**(beetles) Coleoptera**
5. Wings of same length, antennae usually shorter than head 6
   Wings not of same length, antennae long or enlarged toward end.. 7
6. Large insects (usually > 3 cm), wings long, transparent and with many strong veins; abdomen long and slender .... **(dragonflies) Odonata**
   Smaller insects, wing venation faint, wings extending posterior to the abdomen..............**(termites) Isoptera**
7. Wings covered with fine, opaque scales; tubular, coiled, sucking mouthparts............... **(butterflies, moths) Lepidoptera**
   Wings thin, transparent, and not covered with scales; mandibles well developed............**(ants, bees, wasps) Hymenoptera**

3. Your instructor will demonstrate how to effectively use a fish seine (see fig. 8.9).
4. Remember these tips for seining:
   a. Sweep the seine net upstream rather than downstream.
   b. Hold the seine poles so that you push the bottom end in front of you rather than walking backwards and pulling the seine.
   c. Be safe. Move quickly but do not lose control or overwhelm your seining partner.
   d. While moving with your partner, separate the two poles only about one-half to two-thirds the total length of the seine. For example, the poles of a 12-ft seine should be keep 6–9 ft apart, no more.
   e. Keep the lower, weighted edge of the seine against the sediment. As you move, bump the ends of the poles along the bottom of the sediment. This will keep the seine low in the water column.
   f. To finish a seine haul, sweep the net toward and onto the shore rather than lifting the net out of the water while you stand in the stream.
   g. Handle the fish as little as possible. Return all fish to the water alive.
5. Have a team of students waiting to quickly count the fish as soon as the net is brought to the shore.
6. Take two or three seine hauls (totaling > 200 fish) for each sampling site.
7. Count and record in table 12.6 the number of each type (species) of fish for each seine haul.
8. Sum the counts for each species across all seine hauls and record as *Totals* in table 12.6.
9. Calculate and record in table 12.6 richness, Shannon-Wiener diversity, H_{max}, evenness, and Simpson's index for each of the two sampling sites.

Questions 7

What was the most abundant species of fish in your samples? ____

____

Which sampling site was the most diverse? ______________

____

12–9

Did all measures of diversity lead you to the same conclusions? ______________________________

____

Consider the variation among your seine net samples. Would more samples likely lead you to different conclusions? Why? ______________________________

____

Ecologists sometimes express the number of organisms as "catch per-unit-effort" rather than as absolute density. What was the mean catch per-unit-effort for your seine net samples for each sampling site? ______________________________

____

Table 12.6

Fish abundances, classification, and diversity from samples from two contrasting aquatic sites

Site ______						Site ______					
Species	Number in Seine 1	Number in Seine 2	Number in Seine 3	Number in Seine 4	Totals	Species	Number in Seine 1	Number in Seine 2	Number in Seine 3	Number in Seine 4	Totals
Richness ______	Shannon-Wiener ______	H_{max} ______	Evenness ______	Simpson's ______		Richness ______	Shannon-Wiener ______	H_{max} ______	Evenness ______	Simpson's ______	

Questions for Further Thought and Study

1. Which common diversity index would a good ecologist use to characterize a community? How do you justify your answer?

2. How would you design artificial data sets to determine the theoretical maximum and minimum values for Simpson's index?

3. Re-examine the terrestrial habitat that had the highest insect diversity, and the aquatic habitat that had the highest fish diversity. What characteristics of those habitats likely promote diversity?

13

Primary Productivity in an Aquatic Community

Objectives

As you complete this lab exercise you will:

1. Collect lake water samples for analysis of dissolved oxygen, chlorophyll content, and primary production.
2. Measure dissolved oxygen content using the Winkler titration method.
3. Extract chlorophyll from a lake water sample and measure its concentration.
4. Use the light bottle-dark bottle oxygen method to measure primary productivity in a lake.

Communities are alive. They grow, reproduce, and respond to their environment, and it all requires energy. This energy enters the ecosystem as sunlight captured by photosynthesis, which converts inorganic carbon (CO_2) to organic sugars ($C_6H_{12}O_6$). These organic molecules store the energy of sunlight in their carbon bonds. Food chains process the organic molecules and pass the energy from one trophic level to the next (fig. 13.1).

The total organic material synthesized by autotrophs and heterotrophs is called **production** (grams) or **productivity** (grams per unit time), and the organisms that do it are **producers**. Autotrophic organisms are **primary producers** and use sunlight, water (H_2O), and inorganic carbon (CO_2) to synthesize organic molecules. Primary producers include plants and some microorganisms (primarily algae). Heterotrophic organisms are **consumers** and they also produce tissue that is available to the next trophic level. They eat organic molecules and use their energy and building blocks to synthesize their own tissue. This tissue is secondary production and consumers may also be considered **secondary producers**.

The total energy fixed in organic molecules by autotrophs is **gross primary production** (GPP). Part of GPP is metabolized in respiration and part is used for production of tissue (growth) available to the next trophic level (herbivores). The portion of GPP available to herbivores is **net primary production** (NPP).

gross primary production = respiration + net primary production

NPP is either consumed by herbivores or decomposed.

Question 1

Could gross primary production be less than or equal to net primary production? In other words, could all of GPP go towards growth? Why or why not? ______

______

The first step to understand ecosystem energetics is to measure photosynthesis by the community's producers (plants and autotrophic microorganisms). A review of the summary equation for photosynthesis reveals that measuring production can be based on the rate of CO_2 uptake, the increase in weight of tissue (synthesized organic molecules), or the rate of oxygen liberation.

Photosynthesis Summary Equation

$$6\,CO_2 + 12\,H_2O \xrightarrow[\text{chlorophyll}]{\text{light}} C_6H_{12}O_6 + 6\,H_2O + 6\,O_2$$

In this lab exercise you will develop techniques to measure O_2 production and assess primary productivity. You'll also measure chlorophyll content as an index of productivity. Specifically, you will assess primary productivity by measuring (1) dissolved oxygen (DO) using Winkler titration; (2) chlorophyll concentrations by filtering lake water and extracting with acetone; and (3) dissolved oxygen changes due to respiration and photosynthesis in a sample of a freshwater plankton community incubating in sunlight.

Questions 2

How does chlorophyll content generally relate to biomass of primary producers? ______

______

How does chlorophyll content relate to primary productivity?

______

______

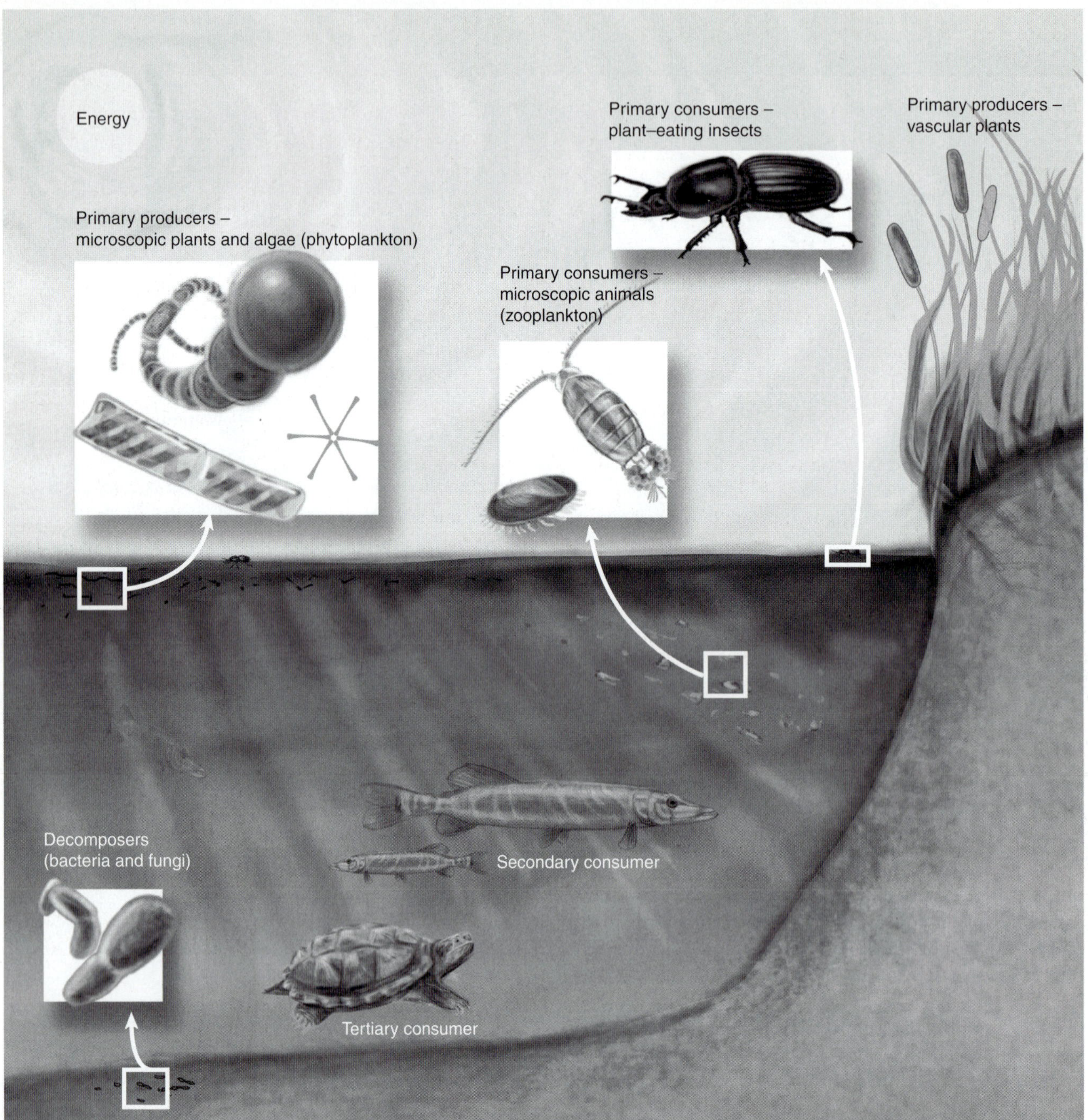

Figure 13.1

Microscopic autotrophs (photosynthetic plants and algae) are primary producers. Their productivity (organic molecules with stored energy) serves as food for primary, secondary, and tertiary consumers. As the food (production) passes through the trophic levels, some is lost to respiration. A portion of the production from each trophic level moves to decomposers.

MEASURING DISSOLVED OXYGEN

Dissolved oxygen (DO) in aquatic ecosystems comes from diffusion of oxygen (O_2) from the atmosphere and O_2 released as a byproduct of photosynthesis. Tracking daily changes in DO is a common technique to assess the health and primary productivity of lakes and streams. DO levels are particularly critical because aquatic life depends on oxygen, and oxygen doesn't dissolve well in water. Atmospheric oxygen (O_2) occurs as 200,000 parts per million (ppm) in air. In contrast, oxygen dissolved in water is commonly 8–10 ppm. This is low. DO of 3 ppm (= 3 mg L^{-1}) generally stresses aquatic organisms. Ecologists often monitor DO concentrations because poor solubility in water means that acquiring oxygen is limiting for aquatic organisms.

Question 3

An accidental discharge of organic sewage into a lake or stream will cause the system's dissolved oxygen concentration to plummet. How so? ________________________

____

A common method to measure DO in a water sample is the **Winkler titration method**. Although specialized electronic meters also can measure dissolved oxygen, the titration method is simple and accurate. Winkler titration is an important part of procedures to assess primary productivity.

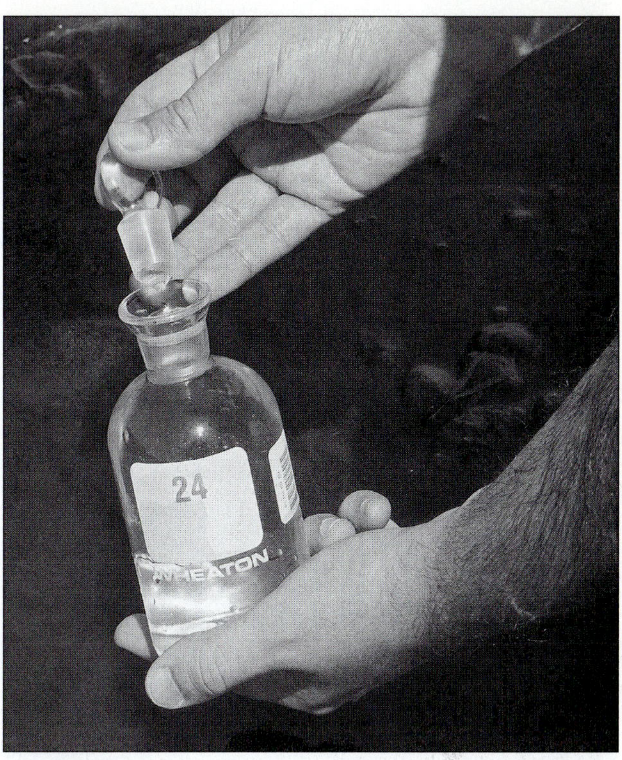

Figure 13.2

This dissolved oxygen bottle will hold a water sample taken from the surface of a lake. A Winkler titration procedure will measure the dissolved oxygen concentration. (See fig. 4.5.)

Procedure 13.1

Practice the Winkler titration method for measuring dissolved oxygen.

1. Examine a DO bottle (fig. 13.2). When it is full, its shape and narrow neck minimizes the surface area exposed to air. Notice that the glass stopper is precisely ground to enclose exactly 300 mL.
2. Fill a 1-L flask with cool tap water, and swirl it vigorously to aerate the water.
3. Fill a DO bottle from the flask and stopper the bottle so no bubbles are trapped. Record in table 13.1 the DO bottle number.
4. Boil the remaining water in the flask for 5 min. This reduces the dissolved gas content.
5. Let the boiled water cool for 10–15 min and use it to fill a second DO bottle. Pour the water gently to minimize mixing with air. Record in table 13.1 the DO bottle number of the boiled sample.
6. Assemble the chemicals, ring stand, and burette provided by your instructor for a Winkler titration.
7. Follow the steps listed in the boxed insert "Winkler Titration: Chemistry and Procedure" in Exercise 4 to determine the DO concentrations in the two water samples. The volume of each DO bottle allows three replicate 100-mL titrations.
8. Record your results in table 13.1.

Question 4

How did boiling affect the dissolved oxygen concentration of the water sample? ________________________

____

Procedure 13.2

Practice collecting an undisturbed lake water sample for dissolved oxygen analysis.

1. Obtain a DO bottle and ground-glass stopper. Rinse the inside of the DO bottle with some lake water.
2. Examine a van Dorn water sampler (see fig. 4.6) and its drain tube. Close the drain valve at the base of the flexible tube.
3. Your instructor will show you how to CAREFULLY set the spring-loaded suction cups, and how to trigger the release mechanism with a weighted messenger (fig. 13.3).
4. Cock the van Dorn sampler. Slowly submerge the cocked van Dorn completely in lake water (or a sink full of water if you are practicing). Lower the van Dorn to the necessary depth (1 m for practice) as marked on the line.

TABLE 13.1

TITRATION VOLUMES AND DISSOLVED OXYGEN CONCENTRATIONS FOR AERATED AND BOILED WATER SAMPLES

	Aerated Water Sample DO Bottle Number _____		Boiled Water Sample DO Bottle Number _____	
	mL of titrant	DO (mg L^{-1})	mL of titrant	DO (mg L^{-1})
Replicate 1	_______	= _______	_______	= _______
Replicate 2	_______	= _______	_______	= _______
Replicate 3	_______	= _______	_______	= _______
		$\bar{x}$ = _______		$\bar{x}$ = _______

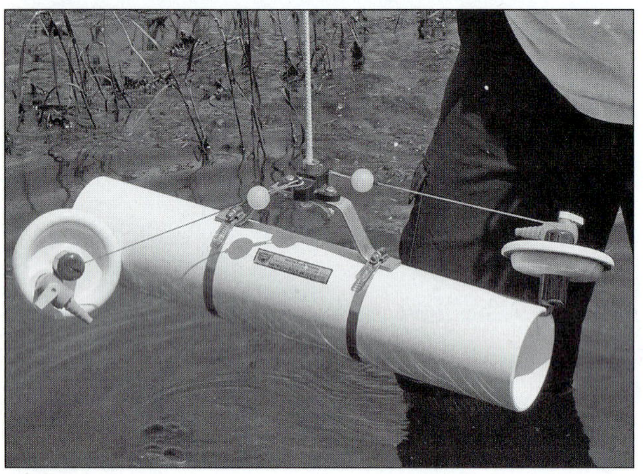

Figure 13.3

This van Dorn water sampler is cocked and ready to be lowered to the desired depth to capture a water sample. The drain tube will be attached after the sampler is brought to the surface. (See Fig. 4.6.)

5. Move the van Dorn about 1 meter horizontally to displace any previously disturbed water in the cylinder.
6. Drop the messenger to close the cylinder and enclose the water sample.
7. Raise the cylinder out of the water and rest it vertically on the edge of a solid surface so the drain valve is at the lower end. Attach the drain tube.
8. Push the drain tube into the DO bottle so the end of the tube touches the bottom of the bottle.
9. Open the drain valve. If water doesn't flow freely into the bottle, lift the edge of the upper suction cup to break the seal and allow air flow.
10. Allow the water to overflow until the volume of the bottle has been displaced three times.
11. As the water continues to flow, slowly pull the tube out of the bottle.
12. Insert the ground-glass stopper into the bottle to seal the 300-mL volume with no bubbles. The sample is now ready for Winkler titration.
13. Empty the van Dorn sampler back into the lake.

Question 5

Why should the volume be overflowed three times while filling a DO sample bottle? ____________________

____

CHOLORPHYLL CONTENT AS A MEASURE OF POTENTIAL PRIMARY PRODUCTIVITY

Chlorophyll captures the energy of sunlight during photosynthesis, and chlorophyll content (mg L^{-1}) is a useful measure of phytoplankton biomass. In turn, phytoplankton biomass is directly proportional to primary productivity in aquatic ecosystems. Although chlorophyll content of the water is not a direct measure of the grams of carbon fixed during photosynthesis, it is a reliable and proportional indicator of primary productivity. Chlorophylls *a* and *b* occur in plants, and a variety of slightly different chlorophylls also occur in algae. Chlorophyll *a* has the broadest occurrence in plants and algae. In the following procedure, phytoplankton is filtered from a water sample and its chlorophyll is extracted with acetone. A spectrophotometer set to chlorophyll's peak absorption wavelength measures the chlorophyll content.

Procedure 13.3

Collect a lake water sample and measure its chlorophyll content.

1. Collect water samples from appropriate depths (surface, 2 m, 4 m, 8 m) using a van Dorn sampler (see Procedure 13.2). Put the samples in labeled, opaque, clean 1-L bottles, and keep them cool until you are ready to filter the water.
2. Assemble a vacuum filtration apparatus (fig. 13.4) compatible with 47-mm diameter membrane filters.
3. Place a 0.8-μm pore size membrane filter on the apparatus and moisten it slightly with a few drops of distilled water.

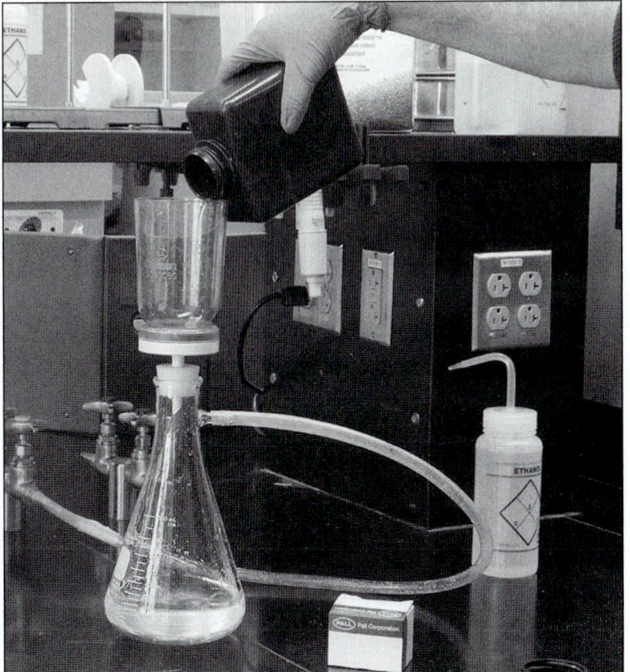

Figure 13.4

In this vacuum filtration apparatus, a small-pore filter has been inserted between the top funnel and bottom collecting flask. A vacuum tube is attached and pulls air from the flask and draws water through the filter to capture suspended phytoplankton.

4. For samples from each depth, shake the bottle to mix any settled algae. Measure 1 L of water from the collected lake sample with a graduated cylinder.
5. Start the vacuum suction and add water from the lake water sample to the filtration receiving funnel.
6. Filter 1 L of water. Continue suction for a few seconds after the last few milliliters have passed through until the filter is damp-dried. If the filter clogs before a full liter is filtered, you can measure the volume of the filtrate and correct the calculations later.
7. Use forceps to carefully remove the filter from the apparatus. Do not touch the filter's upper surface with your fingers.
8. Fold the filter with the phytoplankton on the inside and place it in a labeled 1-in diameter test tube compatible with a spectrophotometer.
9. Add 20 mL of 90% alkalized acetone to the tube and stopper the tube. Shake the tube until the membrane filter dissolves.
10. If the acetone is not slightly green, filter another liter of water and add this second filter to the tube. Record the total volume of water filtered.
11. Repeat steps 3–10 for replicate samples from each depth.
12. Store the tubes overnight in a dark refrigerator to extract all the chlorophyll from the cells.
13. If the solutions are turbid with undissolved material, centrifuge the solutions in labeled, stoppered tubes.
14. Review with your instructor how to use a spectrophotometer. Calibrate the spectrophotometer with an acetone blank.
15. Record each sample and replicate ID in table 13.2. Measure and record in table 13.2 the absorbance of each sample at 750 nm and 663 nm.
16. Calculate the chlorophyll *a* concentration in the extract as:

    chlorophyll *a* (mg L^{-1} extract) = (abs @ 663 nm) − (abs @ 750 nm) × 7.5

    (If the tube light path is 1 in., use 7.5. If the light path is 1 cm, use 13.4.)

17. Calculate the chlorophyll *a* concentration in the lake water sample as:

    chlorophyll *a* (mg L^{-1} lake water) = [chlorophyll *a* (mg L^{-1} extract) × acetone extract volume (mL)] / filtrate volume (L)

18. Your instructor may ask you to graph your results as chlorophyll content per liter of lake water versus depth.

Question 6

Does the chlorophyll content per liter of lake water vary with depth? How so? ______________________________

____

TABLE 13.2

DATA FOR MEASUREMENT OF CHLOROPHYLL CONTENT OF LAKE WATER SAMPLES

Sample Number	Depth	mL Filtered	Absorbance @ 663 nm	Absorbance @ 750 nm	mg Chlorophyll *a* L^{-1} Extract	mg Chlorophyll *a* L^{-1} Lake Water

MEASURING PRIMARY PRODUCTIVITY BY THE LIGHT BOTTLE-DARK BOTTLE OXYGEN METHOD

Primary productivity can be measured by measuring oxygen release during photosynthesis versus oxygen uptake during respiration. For aquatic environments, a common procedure for this is the **light bottle-dark bottle oxygen method.** This method compares changes in DO in lake water samples exposed to light versus samples kept in the dark. Samples of lake water and its phytoplankton are simultaneously incubated in light bottles (clear glass) and dark bottles (opaque) at various depths of the water column.

Both photosynthesis and respiration occur in a light bottle. Only respiration occurs in a dark bottle. During incubation, the initial concentration of dissolved oxygen (DO_{init}) in the dark bottles decreases to DO_{dark} due to respiration. The initial concentration of dissolved oxygen (DO_{init}) in the light bottles increases to DO_{light} as the difference between photosynthetic O_2 release and respiratory O_2 uptake. This assumes that photosynthesis releases more O_2 than respiration consumes.

- ($DO_{init} - DO_{dark}$) measures respiration per unit volume during incubation.
- ($DO_{light} - DO_{init}$) measures net photosynthetic activity.
- The sum of respiration plus net photosynthetic activity $[(DO_{init} - DO_{dark}) + (DO_{light} - DO_{init})]$ measures the gross photosynthetic activity.

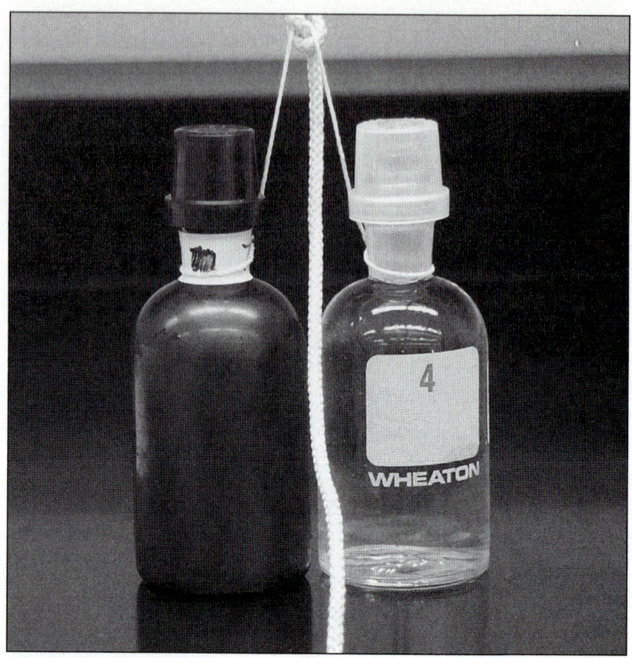

Figure 13.5
This light-dark pair of dissolved oxygen bottles is tied to an incubation line to suspend them in the lake. These and other pairs are tied along the line so the suspended string of bottles will have a pair at each 1-meter depth interval. Productivity decreases as the deeper water decreases light penetration. (See fig. 4.7.)

Procedure 13.4

Prepare light and dark bottles and measure photosynthesis and primary production.

Assemble the incubation line.

1. Discuss with your instructor where to locate a sampling and incubation station in a nearby lake to adequately represent its productivity. Also discuss the number of depths for measuring primary production. One-meter depth intervals are common.
2. Assemble three 300-mL DO bottles for each sampling depth. Cover one of each triplet of bottles completely (light-tight) with tinfoil (fig. 13.5).
3. Assemble some heavy line and a method to attach the bottles to the line so they can be suspended at various depths. Two bottles will be suspended at each depth.

Collect water samples and prepare samples for incubation.

4. Use a van Dorn sampler to sample water from each depth according to Procedure 13.2.
5. For each depth, fill two light bottles and one dark bottle with water from that depth. Use a small square of tinfoil to cover the stopper and neck of the dark bottle. They must be absolutely light tight.
6. Set aside one filled light bottle from each depth for DO_{init} measurement. Attach the other light bottle and the dark bottle to its appropriate depth level on the line so it can later be returned to the water's original depth and incubated there.
7. Record in table 13.3 the identification numbers of the bottles filled from each depth.
8. Suspend the line and the attached pairs of bottles from a flotation device so the line trails into the water and holds each pair of light-dark bottles at the appropriate depth.
9. While in the field, fix the contents of the retained light bottles (DO_{init}) according to the STEPS FOR SAMPLE FIXATION in the boxed reading: Winkler Titration Chemistry and Procedure, Exercise 4.
10. Incubate the samples in the lake for 24 h.

Determine the initial concentration of dissolved oxygen.

11. Return to the lab and titrate the retained light bottles (DO_{init} bottles for each depth) according to the STEPS FOR SAMPLE TITRATION in the boxed reading: Winkler Titration Chemistry and Procedure.
12. Record the titration results for the three 100-mL aliquots from each DO_{init} bottle in table 13.3 as mL of titrant.

Table 13.3

Data and calculations for measurement of primary productivity in lake water by the light bottle–dark bottle oxygen method.

Depth	Bottle ID Number	mL Titrant for 100-mL Aliquot				Dissolved Oxygen
Surface	DO_{init} bottle _____	_____ mL	_____ mL	_____ mL	$\bar{x}$ = _____	= ___ mg L^{-1} = DO_{init}
Surface	DO_{dark} bottle _____	_____ mL	_____ mL	_____ mL	$\bar{x}$ = _____	= ___ mg L^{-1} = DO_{dark}
Surface	DO_{light} bottle _____	_____ mL	_____ mL	_____ mL	$\bar{x}$ = _____	= ___ mg L^{-1} = DO_{light}
Community respiration _____ mg O_2 $L^{-1}d^{-1}$	Net photosynthetic activity _____ mg O_2 $L^{-1}d^{-1}$	Gross photosynthetic activity _____ mg O_2 $L^{-1}d^{-1}$	Net primary productivity _____ mg C $L^{-1}d^{-1}$	Gross primary productivity _____ mg C $L^{-1}d^{-1}$		
1 m	DO_{init} bottle _____	_____ mL	_____ mL	_____ mL	$\bar{x}$ = _____	= ___ mg L^{-1} = DO_{init}
1 m	DO_{dark} bottle _____	_____ mL	_____ mL	_____ mL	$\bar{x}$ = _____	= ___ mg L^{-1} = DO_{dark}
1 m	DO_{light} bottle _____	_____ mL	_____ mL	_____ mL	$\bar{x}$ = _____	= ___ mg L^{-1} = DO_{light}
Community respiration _____ mg O_2 $L^{-1}d^{-1}$	Net photosynthetic activity _____ mg O_2 $L^{-1}d^{-1}$	Gross photosynthetic activity _____ mg O_2 $L^{-1}d^{-1}$	Net primary productivity _____ mg C $L^{-1}d^{-1}$	Gross primary productivity _____ mg C $L^{-1}d^{-1}$		
2 m	DO_{init} Bottle _____	_____ mL	_____ mL	_____ mL	$\bar{x}$ = _____	= ___ mg L^{-1} = DO_{init}
2 m	DO_{dark} Bottle _____	_____ mL	_____ mL	_____ mL	$\bar{x}$ = _____	= ___ mg L^{-1} = DO_{dark}
2 m	DO_{light} Bottle _____	_____ mL	_____ mL	_____ mL	$\bar{x}$ = _____	= ___ mg L^{-1} = DO_{light}
Community respiration _____ mg O_2 $L^{-1}d^{-1}$	Net photosynthetic activity _____ mg O_2 $L^{-1}d^{-1}$	Gross photosynthetic activity _____ mg O_2 $L^{-1}d^{-1}$	Net primary productivity _____ mg C $L^{-1}d^{-1}$	Gross primary productivity _____ mg C $L^{-1}d^{-1}$		
3 m	DO bottle$_{init}$ _____	_____ mL	_____ mL	_____ mL	$\bar{x}$ = _____	= ___ mg L^{-1} = DO_{init}
3 m	DO bottle$_{dark}$ _____	_____ mL	_____ mL	_____ mL	$\bar{x}$ = _____	= ___ mg L^{-1} = DO_{dark}
3 m	DO bottle$_{light}$ _____	_____ mL	_____ mL	_____ mL	$\bar{x}$ = _____	= ___ mg L^{-1} = DO_{light}
Community respiration _____ mg O_2 $L^{-1}d^{-1}$	Net photosynthetic activity _____ mg O_2 $L^{-1}d^{-1}$	Gross photosynthetic activity _____ mg O_2 $L^{-1}d^{-1}$	Net primary productivity _____ mg C $L^{-1}d^{-1}$	Gross primary productivity _____ mg C $L^{-1}d^{-1}$		

Stop the incubation and determine final light and dark dissolved oxygen concentrations.

13. After 24 h incubation, retrieve the light and dark bottles. While in the field, fix the contents of the retrieved bottles according to the STEPS FOR SAMPLE FIXATION in the boxed reading: Winkler Titration Chemistry and Procedure.
14. Return to the lab and titrate the samples according to the STEPS FOR SAMPLE TITRATION in the boxed reading: Winkler Titration Chemistry and Procedure.
15. Record the titration results for the three 100-mL aliquots from each DO_{light} and DO_{dark} bottle in table 13.3 as mL of titrant.

Calculate photosynthetic activity, respiration, and primary production.

16. Calculate and record in table 13.3 the mean ($\bar{x}$) milliliters of titrant per 100 mL for each bottle. This mean value equals the DO concentration in mg L^{-1}.
17. Calculate and record community respiration per day for each depth in table 13.3.

    community respiration (mg O_2 $L^{-1}d^{-1}$) = $DO_{init} - DO_{dark}$

18. Calculate and record net and gross photosynthetic activity for each depth in table 13.3.

    net photosynthetic activity (mg O_2 $L^{-1}d^{-1}$) = $DO_{light} - DO_{init}$

    gross photosynthetic activity (mg O_2 $L^{-1}d^{-1}$) = community respiration + net photosynthetic activity

19. The release of 1 mg O_2 during photosynthesis is equivalent to synthesis of approximately 0.375 mg of carbon in organic molecules. Calculate and record net and gross primary production for each depth in table 13.3.

net primary production (mg C $L^{-1}d^{-1}$) =
net photosynthetic activity × 0.375

gross primary production (mg C $L^{-1}d^{-1}$) =
gross photosynthetic activity × 0.375

20. Your instructor may ask you to graph your results of net and gross primary productivity versus depth.

Questions 7

Does primary productivity differ among the depths sampled? ______________________________

____

Does the depth profile of productivity parallel that of chlorophyll content? Why might they differ? ______________

____

Questions for Further Thought and Study

1. Fish kills often occur in small, nutrient-rich ponds due to oxygen depletion. Why are the dead fish almost always discovered in the morning?

2. How might suspended sediments (turbidity) impact primary productivity?

3. How does primary productivity relate to the number of trophic levels present in an ecosystem?

4. Chlorophyll content is a good indicator of potential primary productivity. However, some algal species compensate for low light conditions by producing more chlorophyll. How would this influence our use of chlorophyll concentration as a predictor of primary productivity?

Competition

14

Objectives

As you complete this lab exercise you will:

1. Investigate the effects of intraspecific competition on individual plant growth, population growth, and age structure of an animal.
2. Experiment with inter- and intraspecific competition pressures on two species of plants.
3. Investigate the effects of plant allelopathic chemicals on germination and success of potentially competing plant species.

Charles Darwin realized that **competition** for limited resources was central to survival, fitness, and evolutionary change. Competition is an interaction among individuals seeking a common resource that is scarce (fig. 14.1)—it negatively impacts the competing organisms. The intensity of competition for a resource depends on the amount of the resource, the number of individuals competing, and the needs of each individual for that resource. Remember that sharing resources such as light, food, water, space, and nutrients is not necessarily competition. There must be a negative effect on the competitors' success. This negative effect usually reduces fitness (reproductive capacity). Competition selects for individuals best adapted to growth and survival in their environment and for those with the greatest reproductive success. This immediate effect of competition may be negative, but long-term competition often leads to partitioning of resources, specialization, and greater community diversity.

Over many generations, natural selection usually dampens the intensity of competition. Competition shifts and separates the overlapping niches of highly competitive species. This separation lessens the negative effects of competition, and it often makes competition difficult to detect among natural populations. For this reason, we often use laboratory populations to demonstrate the negative effects of competition, especially on growth.

Figure 14.1
Light is a valuable resource in a crowded forest. This community's dense growth of plants competes for light and shows a few tall trees forming an upper canopy. A small stream below is almost completely obscured and receives little light to promote algal growth.

In this lab exercise you will measure the effects of competition on growth (1) among sunflower seedlings; (2) between radish and wheat plants; and (3) among beetles growing with limited resources. You will also demonstrate how plants can ease competition for resources by inhibiting the growth of nearby competitors. Competition among members of the same species is **intraspecific competition,** whereas competition among members of different species is **interspecific competition**.

INTRASPECFIC PLANT COMPETITION

Procedure 14.1

Examine competition among sunflower seedlings.

1. Obtain 15 pots containing potting soil.
2. Follow your lab instructor's directions to plant 4, 6, 12, 20, and 40 sunflower seeds with three replicate pots of each treatment. Label each pot with the number of seeds, the date, and your name (fig. 14.2).
3. Water the pots gently with a consistent amount of water.
4. Place the pots randomly in trays in a greenhouse or well-lit area so each pot has the same environmental conditions of light, temperature, etc.
5. After one growth interval (week), remove excess seedlings so the treatments will have 2, 4, 8, 16, and 32 sunflower seedlings.
6. Examine the pots after each of three 1-week intervals. At each interval record general observations and measurements of the parameters called for in table 14.1.
7. After 4 weeks (or a time recommended by your instructor) count and record in table 14.2 the number of plants surviving. Then cut and remove the aboveground tissues of all plants in each pot and place them on a pre-weighed paper towel for each pot.
8. Weigh the paper with plants, subtract the weight of the paper, and record the *Mean fresh weight of aboveground biomass* for the appropriate treatment in table 14.2. Express your results as grams of tissue per pot and as grams of tissue per seed.
9. If you have worked in groups, follow your instructor's directions to either combine and record in table 14.2 the mean and standard error of data from all groups or from your group only.
10. Prepare the following four graphs and draw conclusions. Each graph will have five curves, one for each density of competitors. For each graph draw conclusions about how intraspecific competition affected the variable.
    1. Mean height with standard error bars *vs.* days of growth
       Conclusions: ______________________________

       ____
    2. Mean leaf width with standard error bars *vs.* days of growth
       Conclusions: ______________________________

       ____
    3. Fresh weight (g pot^{-1}) with standard error bars *vs.* days of growth
       Conclusions: ______________________________

       ____

Figure 14.2
Small pots of germinating seeds are replicate sample units for competition experiments. These pots contain sunflower seedlings. Each pot should be marked with a sample ID number.

    4. Fresh weight (g $seed^{-1}$) with standard error bars *vs.* days of growth
       Conclusions: ______________________________

       ____

Questions 1

Was competition greater in the more-crowded pots?

____

____

Which parameters showed the effects of competition?

____

____

What other characteristics of competing plants might you measure for an extended experiment? ______________

____

Did competition more noticeably affect the number of individuals or the biomass of each individual? ______________

____

What kind of environments would likely intensify competition among sunflowers? ______________

____

TABLE 14.1

EFFECTS OF COMPETITION ON SUNFLOWER SEEDLINGS

		Treatment (plants per pot)														
		2			4			8			16			32		
		Rep 1	Rep 2	Rep 3	Rep 1	Rep 2	Rep 3	Rep 1	Rep 2	Rep 3	Rep 1	Rep 2	Rep 3	Rep 1	Rep 2	Rep 3
First interval Total growth days _____	General observations															
Second interval Total growth days _____	General observations															
Third interval Total growth days _____	General observations															
First interval	Mean height of individuals (cm)															
Second interval	Mean height of individuals (cm)															
Third interval	Mean height of individuals (cm)															
First interval	Range of height of individuals (cm)															
Second interval	Range of height of individuals (cm)															
Third interval	Range of height of individuals (cm)															
First interval	Mean width 10 widest leaves (cm)															
Second interval	Mean width 10 widest leaves (cm)															
Third interval	Mean width 10 widest leaves (cm)															

Would you expect different results if different potting soil was used? Why? ______________________________

____

What information might the range of heights provide that the mean height does not provide (review Exercises 1 and 2)? ______________________________

____

Are general observations valuable to your experiment even if they are not quantified? How so? ______________________________

____

What competitive effect of plant density was not tested because you over-planted and then thinned the seedlings to a precise treatment number? ______________________________

____

INTRASPECIFIC ANIMAL COMPETITION

Intraspecific competition reduces growth and fitness, and is best studied in species with short life cycles. Flour beetles (*Tribolium* spp.) (fig. 14.3) are good for laboratory studies of competition because they culture easily and vary in competitive abilities. Be sure to review the egg-larva-pupa-adult life cycle of beetles.

14–3

Table 14.2

Data summary of the effects of competition on sunflower seedlings harvested after four 1-week growth intervals

Total Days of Growth =	Treatment (plants per pot)														
	2			4			8			16			32		
	Rep 1	Rep 2	Rep 3	Rep 1	Rep 2	Rep 3	Rep 1	Rep 2	Rep 3	Rep 1	Rep 2	Rep 3	Rep 1	Rep 2	Rep 3
General observations															
Number plants surviving at harvest															
	Mean =			Mean =			Mean =			Mean =			Mean =		
	Std. err. =			Std. err. =			Std. err. =			Std. err. =			Std. err. =		
	Class mean =			Class mean =			Class mean =			Class mean =			Class mean =		
Mean height of individuals															
	Mean =			Mean =			Mean =			Mean =			Mean =		
	Std. err. =			Std. err. =			Std. err. =			Std. err. =			Std. err. =		
	Class mean =			Class mean =			Class mean =			Class mean =			Class mean =		
Mean width 10 widest leaves															
	Mean =			Mean _=			Mean =			Mean =			Mean =		
	Std. err. =			Std. err. =			Std. err. =			Std. err. =			Std. err. =		
	Class mean =			Class mean =			Class mean =			Class mean =			Class mean =		
Number plants surviving at harvest															
	Mean =			Mean =			Mean =			Mean =			Mean =		
	Std. err. =			Std. err. =			Std. err. =			Std. err. =			Std. err. =		
	Class mean =			Class mean =			Class mean =			Class mean =			Class mean =		
Mean fresh weight of aboveground biomass (g pot^{-1})															
	Mean =			Mean =			Mean =			Mean =			Mean =		
	Std. err. =			Std. err. =			Std. err. =			Std. err. =			Std. err. =		
	Class mean =			Class mean =			Class mean =			Class mean =			Class mean =		
Mean fresh weight of aboveground biomass (g seed^{-1})															
	Mean =			Mean =			Mean =			Mean =			Mean =		
	Std. err. =			Std. err. =			Std. err. =			Std. err. =			Std. err. =		
	Class mean =			Class mean =			Class mean =			Class mean =			Class mean =		

Procedure 14.2

Examine intraspecific competition among *Tribolium*.

1. For each lab group, obtain two pairs (male-female) of *Tribolium confusum*, and a 1-L jar containing 50 g of well-sifted, enriched whole-wheat flour mixed with 1 g of dried yeast.
2. Put the two pairs of beetles in the jar of culture media and cover the jar with a loose, fine-mesh cover.
3. Incubate the culture for two weeks at 25–30°C.
4. Using a fine-mesh screen, separate, count, and record in table 14.3 the number or weight for the total individuals of each life stage. Your instructor will direct you to measure either number or weight, or both.
5. Reintroduce all counted eggs, larvae, pupae, and adults to a jar of fresh culture media.
6. Repeat steps 4 and 5 at 2-week intervals for 2 months. Record your counts in table 14.3.
7. At the conclusion of your experiment, your instructor will direct you to graph your data or the

Figure 14.3
Flour beetles (*Tribolium* sp.) are ideal for experimentation, and they grow well on wheat flour with a minimum of water. Beetles have a distinctive life cycle including larva (top), pupa (middle), and adult (bottom).

combined data from all groups with *time* on the x axis and *number of individuals* on the y axis.

Questions 2

Did high densities of *Tribolium confusum* have competitive consequences? What was your evidence? ______________

____

In what ways other than eating food could *T. confusum* interact competitively? ______________________

____

Did you see any signs of cannibalism? Signs that cannibalism could be a selective force? How so? ______________

____

Did competition for limited resources affect the age structure of the population? How so? ______________

____

Could intraspecific competition be intense enough to eliminate the species? How so? ______________

____

INTRA- AND INTERPSECIFIC PLANT COMPETITION

Procedure 14.3

Examine intraspecific and interspecific competition between plant species.

1. Obtain a supply of radish seeds, wheat seeds, plant labels, and Jiffy® pots. Jiffy pots are dehydrated bags of potting soil commonly sold at garden centers (fig. 14.4).
2. Rehydrate 36 pots by soaking them in water in a large beaker or sink.
3. In separate pots, plant 2, 10, 20, and 40 radish seeds. Make two more replicate treatment sets for a total of 12 pots. Label the pots with IDs as listed in table 14.4.
4. In separate pots, plant 2, 10, 20, and 40 wheat seeds. Make two more replicate treatment sets for a total of 12 pots. Label the pots as listed in table 14.5.
5. In separate pots, plant 1 wheat + 1 radish seed, 5 wheat + 5 radish seeds, 10 wheat + 10 radish seeds, and 20 wheat + 20 radish seeds. Make two more replicate treatment sets for a total of 12 pots. Label the pots as listed in table 14.6.

TABLE 14.3

EFFECTS OF INTRASPECIFIC COMPETITION ON LIFE STAGES OF THE FLOUR BEETLE *TRIBOLIUM CONFUSUM*

	Number or weight of individuals in each life stage				
	Eggs	Larvae	Pupae	Adults	Total Individuals
Days of population growth:	Group: Class:	Group: Class:	Group: Class:	Group: Class:	Group: Class:
Days of population growth:	Group: Class:	Group: Class:	Group: Class:	Group: Class:	Group: Class:
Days of population growth:	Group: Class:	Group: Class:	Group: Class:	Group: Class:	Group: Class:
Days of population growth:	Group: Class:	Group: Class:	Group: Class:	Group: Class:	Group: Class:
Days of population growth:	Group: Class:	Group: Class:	Group: Class:	Group: Class:	Group: Class:
Days of population growth:	Group: Class:	Group: Class:	Group: Class:	Group: Class:	Group: Class:

Figure 14.4

Dry Jiffy pots and pots with radish and wheat seedlings.

6. Place all 36 pots in trays. Put the pots in random positions. Speak to your instructor about how to randomize the pot positions.
7. Allow the seeds to germinate and grow for 10–14 days in a greenhouse.
8. After 10–14 days, count and record in tables 14.4, 14.5, and 14.6 the success of germination (number of plants in each pot).
9. For each pot, harvest the plants and gently shake the soil loose from the roots. Then submerge the roots of the plants in a beaker of water and gently massage away any remaining soil.
10. Blot the plants dry on a paper towel, and place all of the plants from the pot on a pre-dried and pre-weighed paper towel.
11. Weigh the paper with plants, subtract the weight of the paper, and record the net weight (fresh biomass as grams of fresh weight) of the plants in the tables for the appropriate treatments.
12. Air dry the plants and paper towel for 24 h. Reweigh them, and record the dried biomass as grams of air-dried plant in the appropriate tables.
13. Calculate and record in tables 14.4, 14.5, and 14.6 the means of the germination numbers and the fresh and dried weights of each set of replicates.
14. Your instructor may direct you to combine the data from all groups.

TABLE 14.4

GERMINATION RATES AND BIOMASS PRODUCTION BY COMPETING RADISH SEEDLINGS

Treatment	Pot ID	Germination (number of viable plants)	Fresh Biomass (g fresh wt)	Dried Biomass (g air-dry wt)
2 radish seeds				
Rep 1	Rad2-1			
Rep 2	Rad2-2			
Rep 3	Rad2-3			
		Mean =	Mean =	Mean =
10 radish seeds				
Rep 1	Rad10-1			
Rep 2	Rad10-2			
Rep 3	Rad10-3			
		Mean =	Mean =	Mean =
20 radish seeds				
Rep 1	Rad20-1			
Rep 2	Rad20-2			
Rep 3	Rad20-3			
		Mean =	Mean =	Mean =
40 radish seeds				
Rep 1	Rad40-1			
Rep 2	Rad40-2			
Rep 3	Rad40-3			
		Mean =	Mean =	Mean =

Table 14.5

Germination rates and biomass production by competing wheat seedlings

Treatment	Pot ID	Germination (number of viable plants)	Fresh Biomass (g fresh wt)	Dried Biomass (g air-dry wt)
2 wheat seeds				
Rep 1	Whe2-1			
Rep 2	Whe2-2			
Rep 3	Whe2-3			
		Mean =	Mean =	Mean =
10 wheat seeds				
Rep 1	Whe10-1			
Rep 2	Whe10-2			
Rep 3	Whe10-3			
		Mean =	Mean =	Mean =
20 wheat seeds				
Rep 1	Whe20-1			
Rep 2	Whe20-2			
Rep 3	Whe20-3			
		Mean =	Mean =	Mean =
40 wheat seeds				
Rep 1	Whe40-1			
Rep 2	Whe40-2			
Rep 3	Whe40-3			
		Mean =	Mean =	Mean =

Table 14.6

Germination rates and biomass production by competing radish and wheat seedlings

Treatment	Pot ID	Germination (number of viable plants)	Fresh Biomass (g fresh wt)	Dried Biomass (g air-dry wt)
1 radish seed + 1 wheat seed				
Rep 1	RadWhe2-1			
Rep 2	RadWhe2-2			
Rep 3	RadWhe2-3			
		Mean =	Mean =	Mean =
5 radish seeds + 5 wheat seeds				
Rep 1	RadWhe10-1			
Rep 2	RadWhe10-2			
Rep 3	RadWhe10-3			
		Mean =	Mean =	Mean =
10 radish seeds + 10 wheat seeds				
Rep 1	RadWhe20-1			
Rep 2	RadWhe20-2			
Rep 3	RadWhe20-3			
		Mean =	Mean =	Mean =
20 radish seeds + 20 wheat seeds				
Rep 1	RadWhe40-1			
Rep 2	RadWhe40-2			
Rep 3	RadWhe40-3			
		Mean =	Mean =	Mean =

Questions 3

For radish seedlings, which are more significant competitors—wheat seedlings or other radishes? What is your evidence? ______________________________

____

For wheat seedlings, which are more significant competitors—radish seedlings or other wheat? What is your evidence? ______________________________

____

Did interspecific competition reduce the biomass per individual wheat competitor? Radish competitor? __________

____

Did intraspecific competition reduce the biomass per individual wheat competitor? Radish competitor? __________

____

Do your results allow conclusions about fitness of the competitors? Why or why not? ______________________________

____

Did competition more noticeably affect the number of individuals or the biomass of each individual? __________

____

What is your conclusion about the relative intensity of interspecific versus intraspecific competition for radishes and for wheat? ______________________________

____

Would you expect different results if different potting soil was used? Why?______________________________

____

How would you design an experiment to test if radish and wheat compete for space versus nutrients? __________

____

Allelopathy

The "struggle for existence" at the heart of Darwin's model of evolution conjures visions of violent battles among animals vying for scarce resources. But more subtle forms of "combat" are common in animals and plants. One such mechanism of competition is allelopathy. Some plants produce chemicals that inhibit the growth of nearby plants. **Allelopathy** is the inhibition of a plant's germination or growth by exposure to compounds produced by another plant. Allelopathic compounds can be airborne or leach from various plant parts into the soil. Rainfall, runoff, and diffusion distribute inhibitory compounds in the immediate area of the producing plant. In the nearby area, allelopathic compounds inhibit germination, growth, or reproduction of potential competitors (fig. 14.5).

Questions 4

What are adaptive advantages of producing allelopathic compounds? ______________________________

____

What are possible disadvantages of producing allelopathic compounds? ______________________________

____

Procedure 14.4

Demonstrate allelopathy.

1. Determine with your instructor the overall experimental design for the class. Determine how many plant extracts your group will test. This procedure assumes three replicates. Copy table 14.7 for each extract being tested.
2. Obtain tissue (stems and leaves) from the variety of plants provided by your instructor. Some of these plants may produce allelopathic compounds.
3. For each plant, homogenize 10 g of tissue with 100 mL of water in a blender or mortar and pestle. Let the slurry soak for 5–10 min to leach chemicals from the disrupted tissue.

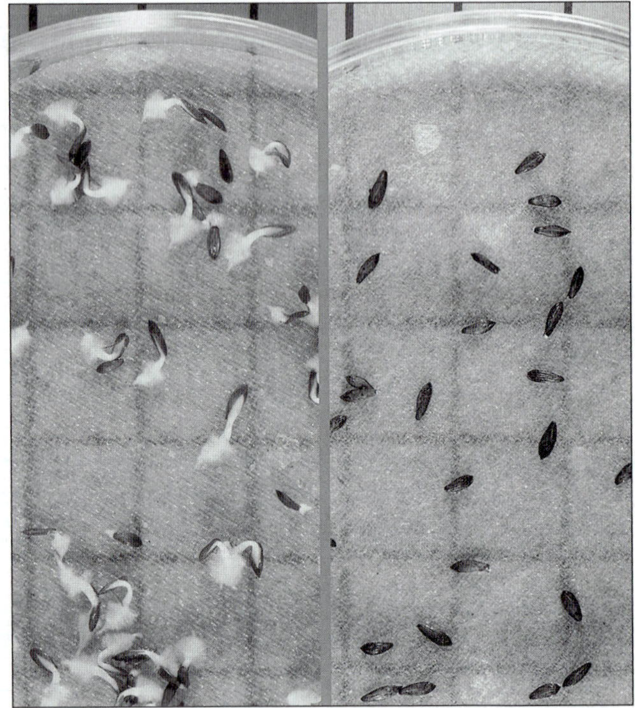

Figure 14.5

Comparison of germination success versus failure of seedlings in an extract with allelopathic chemicals. The lettuce seeds on the right are on a paper towel soaked with an allelopathic chemical. Seeds on the left are controls.

4. Filter or strain the slurry to remove large particulates. Collect the filtrate in a beaker.
5. For each plant extract:
   a. Obtain six petri dishes (three treatment, three control) and line the bottoms with circular pieces of filter paper.
   b. Label each petri dish with the plant extract name and replicate ID number for that dish.
   c. Record in table 14.7 the plant extract name and replicate IDs.
   d. Saturate the filter paper in three of the dishes with 5 mL of the extract. Saturate the filter paper of three dishes with the same amount of water from the same source used to prepare the extract.
   e. Obtain seeds of radish, lettuce, or oat. Distribute 50 seeds uniformly on the filter paper in each dish.
6. Your instructor may extend the experimental design by asking you to set up treatments of multiple extracts and to test the effects on different kinds of seeds. Follow their directions.
7. Incubate the covered dishes at room temperature in the laboratory or in a greenhouse. After 24 and 48 h, count and record in table 14.7 the number of germinated seeds and calculate the percent germination for each replicate and control dish.
8. After 72 h (or the time specified by your instructor), measure the length of 10 radicles randomly subsampled from each dish. Record the lengths in table 14.7.
9. Compare each control mean with the appropriate treatment mean to determine if the extract significantly retarded, enhanced, or had no effect on germination or growth.

Questions 5

Was allelopathy apparent from the tested plant species?

____

____

Which plant species has the most intense allelopathy?

____

____

Why was water used as a comparable treatment? ________

____

How would you detect allelopathy in the field? ________

____

Table 14.7

Data for Germination and Radicle Growth by Seeds Exposed to Potentially Allelopathic Plant Extracts

Plant extract ______			
Petri dish ID	24-h Germination (number of germinated seeds)	48-h Germination (number of germinated seeds)	72-h Radicle Lengths (mm)
Rep1-____			__ __ __ __ __ __ __ __ __ __
Rep2-____			__ __ __ __ __ __ __ __ __ __
Rep3-____			__ __ __ __ __ __ __ __ __ __
	Treatment mean =	Treatment mean =	Treatment mean =
Rep1-water			__ __ __ __ __ __ __ __ __ __
Rep2-water			__ __ __ __ __ __ __ __ __ __
Rep3-water			__ __ __ __ __ __ __ __ __ __
	Control mean =	Control mean =	Control mean =
	Control − treatment mean =	Control − treatment mean =	Control − treatment mean =

AN INVESTIGATION: COMPARE ALLELOPATHIC CHEMICAL PRODUCTION IN ROOTS, STEMS AND LEAVES

Not all organs (i.e., roots, stems, leaves) of allelopathic plants produce equal amounts of allelopathic chemicals.

Procedure 14.5

Compare allelopathy from various plant tissues.

1. Use Procedure 14.4 to document the allelopathic chemical production by a readily available plant species of your choice.
2. Design an experiment to compare allelopathic chemical production by roots, stems, leaves, and flowers of the selected plant.
3. Review Exercises 1 and 2, and form a testable hypothesis about the comparison by your experiment. Write your hypothesis here: ______________________

    ____

4. Describe your experimental design here: __________

    ____

    ____

5. Do your experiment.

Questions 6

Do you accept or reject your hypothesis? ______________

____

What do you conclude about variation in allelopathic chemical production in different plant organs? __________

____

Questions for Further Thought and Study

1. How does competition influence natural selection? Is the presence of competitors a selective force?

2. What characteristics indicate that a community has been undisturbed for a few years? Is there a link between disturbance and the outcome of competition between two species?

3. Why would we expect natural selection to dampen the intensity of competition over many generations?

4. Would you expect inter- or intraspecific competition to be the most intense? Why?

5. Would plants and animals compete for the same resources? How so?

15

Natural Selection

Objectives

As you complete this lab exercise you will:

1. Examine working definitions of *evolution*, *fitness*, *selection pressure*, and *natural selection*.
2. Determine the genotypic and phenotypic frequencies within a population and apply the terms *allele*, *dominant*, *recessive*, *homozygous*, and *heterozygous*.
3. Use the Hardy-Weinberg Principle to demonstrate negative selection pressures on a population.

Species and their environments change with time—without a doubt. To ecologists, the most profound changes are genetic. The **theory of evolution** broadly describes genetic change in populations. Many mechanisms can change the genetic makeup of populations, and our understanding of the relative importance of each mechanism is constantly being refined. Nevertheless, genetic change and, therefore, evolution, are universally accepted by ecologists. Events such as **mutations** (changes in the genetic message of a cell) and catastrophes (e.g., meteor showers, ice ages) all lead to some degree of genetic change. However, all modern evidence points to natural selection as the major force behind genetic change and evolution.

Charles Darwin first described the mechanics of natural selection (fig. 15.1). Darwin postulated that organisms that survive and reproduce successfully in a competitive environment must have traits better adapted for their environment than those of their competitors. In other words, adaptive traits increase organisms' fitness, and these traits are passed more frequently to the next generation. If traits of the most fit individuals are transmitted to the next generation through increased reproduction, then the frequency of these traits will, after many generations, increase in the population. Subsequently, the population and its characteristics will gradually change. Darwin called this overall process **natural selection** and proposed it as a major force guiding genetic change and the formation of new species.

Figure 15.1

Darwin greets his "monkey ancestor." In his time, Darwin was often portrayed unsympathetically, as in this drawing from an 1874 publication.

Review in your textbook the theories of evolution and the mechanism of natural selection.

Natural selection in living populations over many generations is difficult to demonstrate in the lab. Therefore, in this exercise you will *simulate* reproducing populations with nonliving, colored beads representing organisms and their gametes. This artificial population quickly reveals genetic change over many generations. Before you begin, review the terms **gene**, **allele**, **dominant alleles**, **recessive alleles**, **homozygous**, and **heterozygous**.

You will begin your experiments with a "stock population" of organisms consisting of a container of beads. Each bead represents a *haploid* gamete (having one set of chromosomes). Its color represents the allele it is carrying. An organism from this population is *diploid* (has two sets of chromosomes per nucleus) and is represented by two beads.

UNDERSTANDING ALLELIC AND GENOTYPIC FREQUENCIES

Frequency refers to the proportion of alleles, genotypes, or phenotypes of a certain type relative to the total number considered. Frequency is a decimal proportion of the total alleles or genotypes in a population. For example, if 1/4 of the individuals of a population are genotype *Bb*, the genotypic frequency of *Bb* is 0.25. If 3/4 of all alleles in a population are *B*, then the frequency of *B* is 0.75. Remember, by definition the frequencies of all possible alleles or genotypes or phenotypes will always total 1.0.

In the following procedures you will simulate evolutionary changes in allelic and genotypic frequencies in an artificial population.

- The trait is fur color.
- A colored bead is a gamete with a dominant allele (complete dominance) for black fur (*B*)
- A white bead is a gamete with a recessive allele for white fur (*b*).
- An individual is represented by two gametes (beads).
- Individuals with genotypes *BB* and *Bb* have black fur and those with *bb* have white fur.

Procedure 15.1

Establish a parental population.

1. Obtain a "stock population" of organisms consisting of a container of colored and white beads.
2. Obtain an empty container marked "Parental Population."
3. From the stock population select nine homozygous dominant individuals (*BB*) and place them in the container marked "Parental Population." Each individual is represented by two colored beads.
4. From the stock population select 42 heterozygous individuals (*Bb*) and put them in the container marked "Parental Population." Each individual is represented by a colored and a white bead.
5. From the stock population select 49 homozygous recessive individuals (*bb*) and put them in the container marked "Parental Population." Each individual is represented by two white beads.
6. Calculate the total number of individuals and the total number of alleles in your newly established parental population. Use this information to calculate and record in table 15.1 the correct genotypic frequencies for your parental population.
7. Complete table 15.1 with the number and frequency of each of the two alleles.

Questions 1

How many of the total beads are colored? ____________

How many are white? ____________

What color of fur do *Bb* individuals have? ____________

____

How many beads represent the population of 100 organisms? ____________

____

THE HARDY-WEINBERG PRINCIPLE

The **Hardy-Weinberg Principle** enables us to calculate and predict allelic and genotypic frequencies. We can compare these predictions with actual changes that we observe in natural populations and learn about factors that influence gene frequencies.

This predictive model includes two simple equations first described for stable populations by G. H. Hardy and W. Weinberg. Hardy-Weinberg equations (1) predict allelic and genotypic frequencies based on data for only one or two frequencies; and (2) establish theoretical gene frequencies that we can compare to frequencies from natural populations. For example, if we know the frequencies of *B* and *BB*, we can use the Hardy-Weinberg equations to calculate the frequencies of *b*, *Bb*, and *bb*. Then we can compare these frequencies with those of a natural population that we might be studying. If we find variation from our predictions, we can study the reasons for this genetic change.

For the Hardy-Weinberg equations, the frequency of the dominant allele of a pair is represented by the letter p, and that of the recessive allele by the letter q. Also, the genotypic frequencies of *BB* (homozygous dominant), *Bb* (heterozygous), and *bb* (homozygous recessive) are represented by p^2, $2\,pq$, and q^2, respectively. Examine the frequencies in

TABLE 15.1

FREQUENCIES OF GENOTYPES AND ALLELES OF THE PARENTAL POPULATION

Genotypes	Frequency
BB ●●	
Bb ●○	
bb ○○	

Alleles	Frequency
B ●	
b ○	

Phenotype	Frequency
Black fur	
White fur	

table 15.1 and verify calculations of the Hardy-Weinberg equations:

$$p + q = 1$$

$$p^2 + 2\,pq + q^2 = 1$$

The Hardy-Weinberg Principle and its equations predict that frequencies of alleles and genotypes remain constant from generation to generation in stable populations. Therefore, these equations can be used to predict genetic frequencies through time. However, the Hardy-Weinberg prediction assumes that:

- The population is large enough to overcome random events.
- Choice of mates is random.
- Mutations do not occur.
- Individuals do not migrate into or out of the population.
- Natural or artificial selection pressures are not acting on the population.

Questions 2

Consider the Hardy-Weinberg equations. If the frequency of a recessive allele is 0.3, what is the frequency of the dominant allele? ______________________________

____

If the frequency of the homozygous dominant genotype is 0.49, what is the frequency of the dominant allele? ______

____

If the frequency of the homozygous dominant genotype is 0.49, what is the frequency of the homozygous recessive genotype? ______________________________

____

Which Hardy-Weinberg equation relates the frequencies of the alleles at a particular gene locus? ______________

____

Which Hardy-Weinberg equation relates the frequencies of the genotypes for a particular gene locus? ____________

____

Which Hardy-Weinberg equation relates the frequencies of the phenotypes for a gene? ____________________

____

To verify the predictions of the Hardy-Weinberg Principle, use the following procedure to produce a generation of offspring from the parental population you created in the previous procedure. Remember, the fact that the genetic frequencies of various alleles, genotypes, and phenotypes total 1.0 is not a prediction of the Hardy-Weinberg Principle. The total of 1.0 is a mathematical fact. The prediction is that the relative frequencies will not change if all assumptions are met.

Procedure 15.2

Verify the Hardy-Weinberg Principle.

1. Examine figure 15.2 for an overview of the steps of this procedure.
2. Establish the parental population described in Procedure 15.1 (fig. 15.2*a*, 15.2*b*).
3. Simulate random mating of individuals by mixing the population (fig. 15.2*c*).
4. Reach into the parental container (without looking) and randomly select two gametes. Determine their genotype (fig. 15.2*d*).
5. Record the occurrence of the offspring's genotype in figure 15.2*e* as a mark under the heading "Number," or temporarily on a second sheet of paper and return the beads to the container.
6. Repeat steps 4 and 5 (100 times) to simulate the production of 100 offspring.
7. Calculate the frequency of each genotype and allele, and record the frequencies in figure 15.2*e*. Beside each of these new-generation frequencies write (in parentheses) the original frequency of that specific genotype or allele from table 15.1.

Questions 3

The Hardy-Weinberg Principle predicts that genotypic frequencies of offspring will be the same as those of the parental generation. Were they the same in your simulation?

____

____

If the frequencies were different, then one of the assumptions of the Hardy-Weinberg Principle was probably violated. Which one? ______________________________

____

EFFECT OF A SELECTION PRESSURE

Selection is the differential reproduction of phenotypes—that is, some phenotypes (and their associated genes) are passed to the next generation more often than others. In positive selection, genotypes representing adaptive traits in an environment increase in frequency because their bearers survive and reproduce more. In negative selection, genotypes representing nonadaptive traits in an environment decrease in frequency because their bearers are less likely to survive and reproduce.

Selection pressures are factors such as temperature and predation that result in selective reproduction of phenotypes. Some pressures may elicit 100% negative selection against a characteristic and eliminate *all* successful reproduction by individuals having that characteristic. For example, mice with white fur may be easy prey for a fox if they live on a black lava field. This dark environment is a

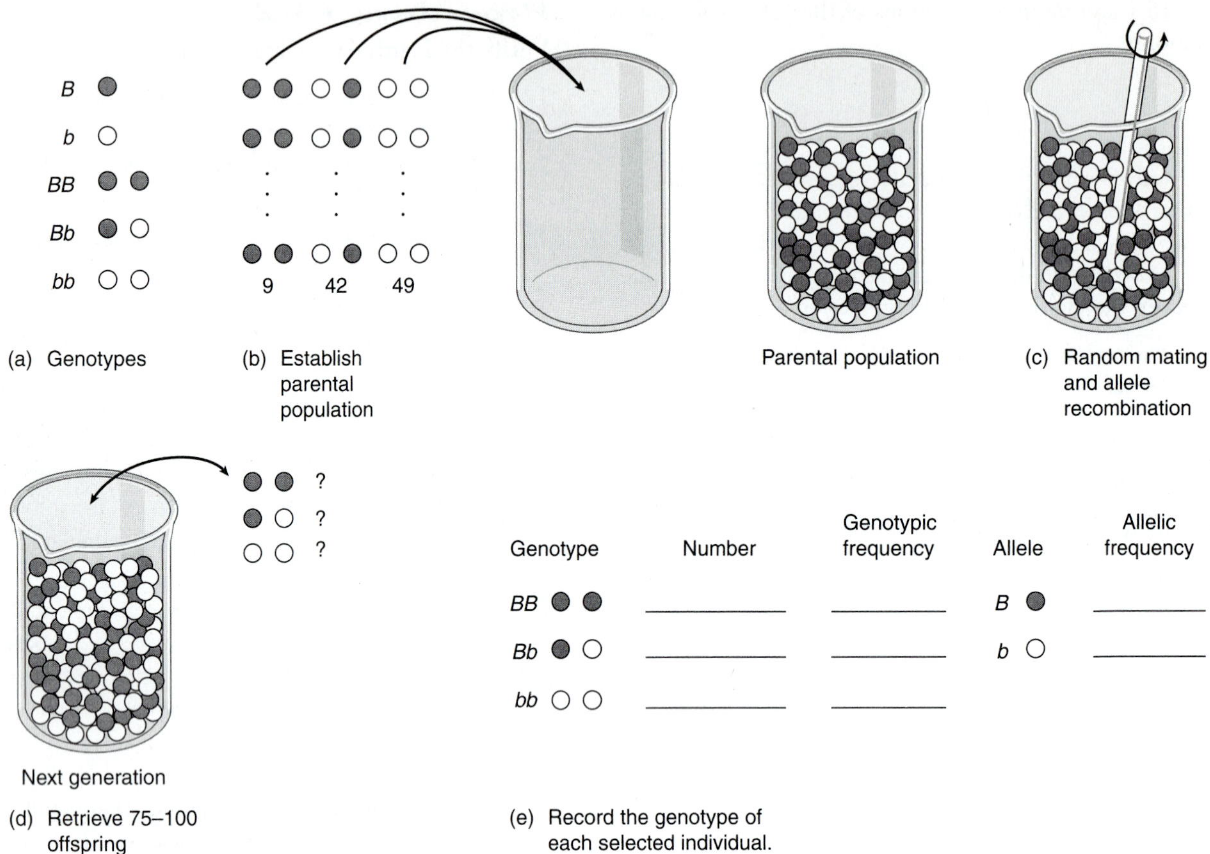

Figure 15.2
Steps in the verification of the Hardy-Weinberg Principle.

negative selection pressure against white fur. If survival and reproduction of mice with white fur were eliminated (i.e., if there is 100% negative selection), would the frequency of white mice in the population decrease with subsequent generations? To test this, use the following procedure to randomly mate members of the original parental population to produce 100 offspring (fig. 15.3).

Procedure 15.3

Simulate 100% negative selection pressure.

1. Establish the same parental population (Procedure 15.1) you used to test the Hardy-Weinberg prediction.
2. Simulate the production of an offspring from this population by randomly withdrawing two gametes to represent an individual offspring (fig. 15.3).
3. If the offspring is *BB* or *Bb*, place it in a container for the accumulation of the "Next Generation." Record the occurrence of this genotype on a separate sheet of paper.
4. If the offspring is *bb*, place this individual in a container for those that "Cannot Reproduce." Individuals in this container should not be used to produce subsequent generations. Record the occurrence of this genotype on a sheet of paper.
5. Repeat steps 2–4 until the parental population is depleted, thus completing the first generation.
6. Calculate the frequencies of each of the three genotypes recorded on the separate sheet and record these frequencies for the first generation in table 15.2. Individuals in each generation will serve as the parental population for each subsequent generation.
7. Repeat steps 2–5 to produce second, third, fourth, and fifth generations. After the production of each generation, record your results in table 15.2.
8. Graph your data from table 15.2 using the graph paper at the end of this exercise. *Generation* is the independent variable on the x axis and *Genotype* is the dependent variable on the y axis. Graph three curves, one for each genotype.

Because some members of each generation (i.e., the *bb* that you removed) cannot reproduce, the number of offspring from each successive generation of your population will decrease. However, the frequency of each genotype, not the number of offspring, is the important value.

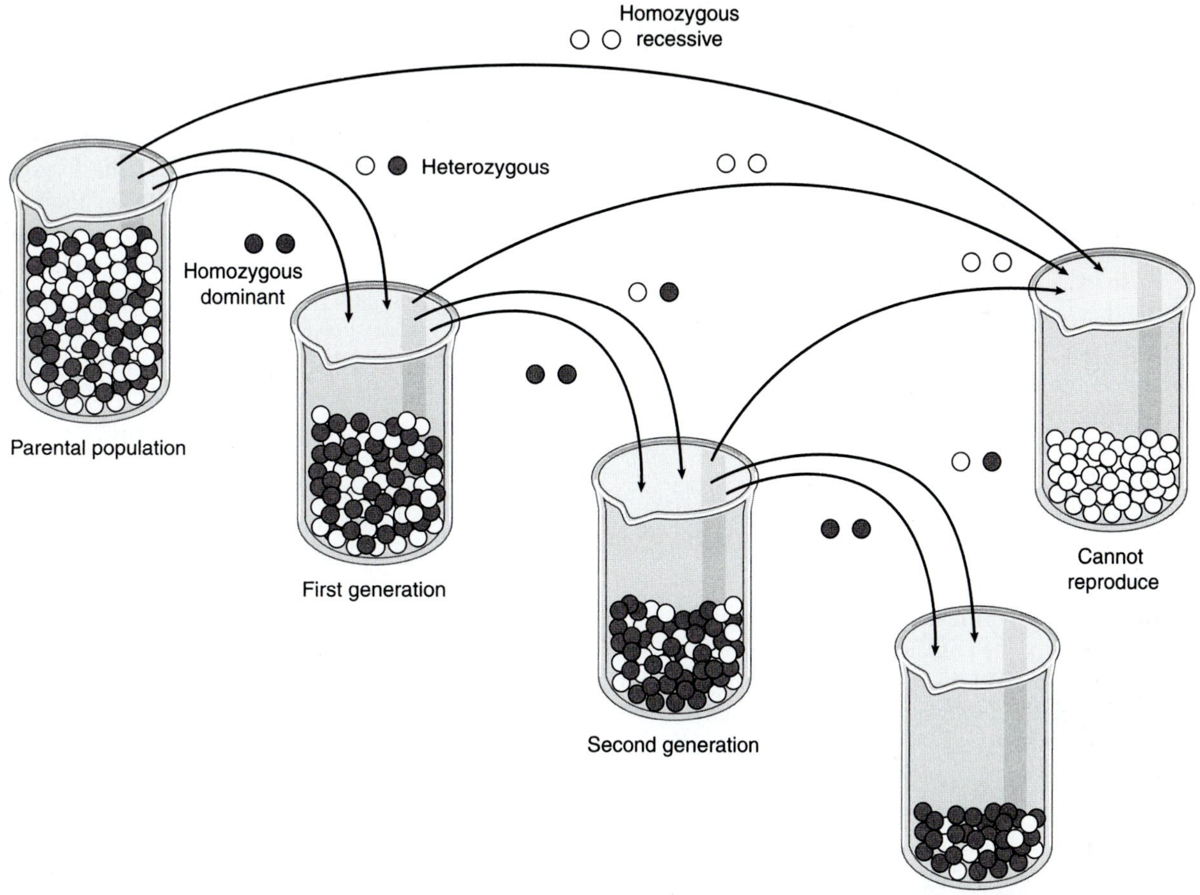

Figure 15.3

Demonstrating the effect of 100% selection pressure on genotypic and phenotypic frequencies across three generations. Selection is against the homozygous recessive genotype. Random mating within the parental population is simulated by mixing the gametes (beads), and the parental population is sampled by removing two alleles (i.e., one individual) and placing them in the next generation. Homozygous recessive individuals are removed (selected against) from the population. The genotypic and phenotypic frequencies are recorded after the production of each generation. The production of each generation depletes the beads in the previous generation in this simulation.

Table 15.2

Genotypic Frequencies for 100% Negative Selection

	Generation				
Genotype	First	Second	Third	Fourth	Fifth
BB ●●					
Bb ●○					
bb ○○					
Total	1.0	1.0	1.0	1.0	1.0

Questions 4

Did the frequency of white individuals decrease with successive generations? Explain your answer. ________________

____

Was the decrease of white individuals from the first to second generation the same as the decrease from the second to the third generation? From the third to the fourth generation? Why or why not? ________________

____

How many generations would be necessary to eliminate the allele for white fur? ________________

____

15–5

Most natural selective pressures do not completely eliminate reproduction by the affected individuals. Instead, their reproductive capacity is reduced by a small proportion. To show this, use Procedure 15.4 to eliminate only 20% of the *bb* offspring from the reproducing population.

Procedure 15.4

Simulate 20% negative selection pressure.

1. Establish the same parental population (Procedure 15.1) that you used to test the Hardy-Weinberg prediction.
2. Simulate the production of an offspring from this population by randomly withdrawing two gametes to represent an individual offspring.
3. If the offspring is *BB* or *Bb*, place it in a container for production of the "Next Generation." Record the occurrence of this genotype on a separate sheet of paper.
4. If the offspring is *bb*, place every fifth individual (20%) in a separate container for those that "Cannot Reproduce." Individuals in this container should not be used to produce subsequent generations. Place the other 80% of the homozygous recessives in the container for the "Next Generation." Record the occurrence of this genotype on a sheet of paper.
5. Repeat steps 2–4 until the parental population is depleted, thus completing the first generation.
6. Calculate the frequencies of each of the three genotypes recorded on the separate sheet and record these frequencies for the first generation in table 15.3.
7. Repeat steps 2–5 to produce second, third, fourth, and fifth generations. Individuals in the "Next Generation" serve as the parental population for each subsequent generation. After production of each generation, record your results in table 15.3.
8. Graph your data from table 15.3 using the graph paper at the end of this exercise. *Generation* is the independent variable on the x axis and *Genotype frequency* is the dependent variable on the y axis. Graph three curves, one for each genotype.

Because some members of each generation cannot reproduce, the number of offspring from each generation of your population will decrease. However, the frequency of each genotype, not the number of offspring, is the important value.

Questions 5

Did the frequency of white individuals decrease with successive generations? ______________________________

____

Was the rate of decrease for 20% negative selection similar to the rate for 100% negative selection? If not, how did the rates differ? ______________________________

____

Table 15.3

Genotypic frequencies for 20% negative selection

	Generation				
Genotype	First	Second	Third	Fourth	Fifth
BB ●●					
Bb ●○					
bb ○○					
Total	1.0	1.0	1.0	1.0	1.0

Questions for Further Thought and Study

1. Charles Darwin wasn't the first person to suggest that populations evolve, but he was the first to describe a credible mechanism for the process. That mechanism is natural selection. What is natural selection? How can natural selection drive evolution?

2. How would selection against heterozygous individuals over many generations affect the frequencies of homozygous individuals? Would the results of such selection depend on the initial frequencies of p and q? Could you test this experimentally? How?

3. How are the frequencies of genes for nonreproductive activities such as feeding affected by natural selection?

4. Do you suspect that evolutionary change always leads to greater complexity? Why or why not?

5. Is natural selection the only mechanism of evolution? Explain.

6. What change in a population would you expect if a selection pressure was against the trait of the dominant allele?

15–8

16

Adaptations of Vertebrates to Their Environment

Objectives

As you complete this lab exercise you will:

1. Examine skeletal adaptations of vertebrates.
2. Recognize the functions of external adaptations of vertebrates that contribute to fitness.
3. Simulate competition and success among similar morphological adaptations.

Adaptations are characteristics and structures of an organism that facilitate vital processes such as homeostasis, food getting, and reproduction. Not all characteristics of an organism necessarily promote fitness. For example, a vertebrate's chin or claw color may be a neutral structural necessity or a by-product of other features. However, characteristics that promote survival and reproduction and are subject to environmental selection are considered adaptations. They promote fitness. They result from a species' long-term interaction with its environment and are shaped by natural selective pressures that promote or retard the passing of the genetic blueprints of an adaptation to the next generation. Examining structural adaptations quickly reveals that many serve multiple functions and are best studied in the context of their environment.

In this lab exercise you will examine morphological characteristics common to groups of animals well-adapted to their environment. As you examine each adaptation, consider the kind of environment that promotes and selectively hones the gene frequencies of the adaptation's genetic blueprint. Adaptations relate directly to the ecology of a species.

ADAPTIVE SKELETAL FEATURES

Primary among the many functions of the skeletal system is providing sites for muscle attachment for flexible movement. Movement, especially locomotion, involves generating force and overcoming gravity. This requires rigid bones to resist powerful muscles, and flexile joints for coordinated movement. Skeletal adaptations are remarkably varied.

Procedure 16.1

Examine skeletal adaptations of representatives from the major classes of vertebrates.

1. Examine skeletons representing the major classes of vertebrates.
2. Consider the environment of each organism. The environment selects for efficient functional morphology needed for locomotion.

Questions 1

Undulation is efficient in water and requires a flexible axial skeleton. What part of a fish's skeleton provides for flexibility needed for undulatory swimming? ________________

____

What percentage of a fish's length includes flexible vertebrae? ________________________________

Are there other ways of moving through an aquatic environment besides undulation? How so? ________________

____

____

3. The ecology and life history of amphibians is associated with an aquatic environment. Examine a skeleton of *Necturus* (mud puppy).

Questions 2

Is an amphibian such as a mud puppy adapted for swimming? Crawling? ________________________

____

What does this tell you about the microenvironment occupied by such amphibians? ____________________

____

How does length and flexibility of the vertebral column compare with pectoral and pelvic appendage development? Which is more robustly developed—the vertebral column or appendages? ________________________

____

Which form of locomotion likely generates the most power for a mud puppy? What is your evidence? ______________

____

Frogs are also amphibians associated with water. What percentage of their body length includes flexible vertebrae? Why so small? ______________

____

What do you conclude about the environment that has shaped frogs' adaptations for locomotion? ______________

____

Which frog bones appear best adapted for attachment of powerful muscles? ______________

____

Frogs have long digits. Are they adaptive for a strong grip? If not, what is the adaptive value of long digits for frogs?

____

____

4. Reptiles include transitional morphologies adapted to life on land. Examine skeletons of a turtle, a snake, and a small alligator.

Questions 3

Is undulation (versus crawling or walking) a viable form of locomotion in a terrestrial environment? What is your evidence? ______________

____

Terrestrial environments lack the buoyancy of water. Powerful muscles are needed to lift and move body mass. Are appendages of reptiles more developed than those of fish and amphibians? ______________

____

The axial skeleton includes the head and spinal column. Which of the reptile skeletons available has the least axial flexibility? ______________

____

Does a lack of axial flexibility mean that the skeleton is not well-adapted to its environment? ______________

____

How does the anatomy and ecology of the organism compensate for less undulation to power locomotion? ______________

____

Which features of an alligator's developed appendages and axial flexibility are adaptive for its methods of locomotion?

____

____

Has natural selection produced a singular "best" morphology for locomotion? Why or why not? ______________

____

Are there flying reptiles? Have there ever been? ______________

____

5. Examine a bird skeleton.

Questions 4

Birds share a recent and direct lineage to reptiles. How much axial flexibility for locomotion does a bird skeleton have? ______________

____

Fish and amphibians and some reptiles use extended digits to push against their environment as they move. Are the digits of a bird's wing well-developed? If not, what other adaptations accomplish this same function? ______________

____

Land animals must compensate for gravity with strong muscles and appendages. Birds, however, must not only compensate, but must overcome gravity. This takes powerful muscles. What adaptive skeletal feature provides for broad attachment of powerful muscles? ______________

____

The rigors of land and air environments select for powerful and adaptive bones and muscles. Which bones of the bird are the thickest and most robust? ______________

____

6. Examine a cat skeleton.

Questions 5

Which are among the most vital organs of mammals and birds? ______________

____

What skeletal structures are adapted to protect these organs? ______________

____

Which bones of the cat are the thickest and most robust?

____

____

Question 6

Are the bones of a fish as thick and robust as those of a cat? Why or why not? ______________

____

7. Compare the teeth of all the vertebrate skeletons available, including fish, frog, alligator, bird, cat, and other mammals.

Questions 7

From your experience, are fish, amphibians, and reptiles "gulpers" or "chewers" when they eat? ______________

____

What is your evidence for gulping or chewing from their skeletal morphology? ______________

____

Do the teeth of an alligator have much variation? Or are they all about the same length and shape? What are they adapted to do? ______________

____

Which of the vertebrates on display show marked variation between front teeth and cheek teeth? ______________

____

Some mammals have cheek teeth adapted for grinding and some have cheek teeth for cutting. How are a cat's cheek teeth adapted? A human's cheek teeth? Horse? Cow?

____

____

ADAPTIVE EXTERNAL FEATURES

Natural selection has shaped available genetic variation and the results are adaptations. Over many generations, characteristics with no adaptive advantage for survival and reproduction may decrease in frequency and those with significant advantage become prominent and frequent. Adaptive external features are an organism's interface with its environment and are subject to strong selective pressures. External adaptations and their functions vary a great deal among the classes of vertebrates. Among the major functions subject to selective pressures are:

- Protection
- Sensing the environment
- Locomotion
- Gas exchange

Question 8

Do you expect some external features to serve more than one adaptive function? For example? ______________

____

Procedure 16.2

Examine adaptations of the external features of representatives from the major classes of vertebrates.

1. Examine table 16.1 and the four broad functions listed. Can you add to the list?
2. Examine the external features of each of the major classes of vertebrates.
3. Identify as many external features as possible for each specimen. Make a note in table 16.1 concerning the function(s) for which each feature confers an advantage.

A SIMULATION AND TEST OF ADAPTIVE MORPHOLOGIES

A widely studied example of subtle variation of an adaptation involves the beaks and feeding ecology of Darwin's finches of the Galápagos Islands. Review this topic. When the parent population of finches arrived on the Galápagos, the birds became isolated as subpopulations on the islands. With time, speciation occurred and subpopulations evolved beaks adapted to particular food items in the varied island environments. Food availability and competition were selective pressures that shaped beak morphologies, allowing each species to exploit a particular food.

In the following procedure each student in a team of four has a different hand tool analogous to the beak of a feeding bird. That beak represents an adaptation to gather food items of a particular size or shape. Some adaptations (beaks) are more advantageous than others at gathering food of a particular size. In a competitive environment, the organism with the best adaptive morphologies will gather more food and will therefore be more fit. The four students will simultaneously feed from the same resource, and their success at gathering food will measure the effectiveness of the "beak" adaptations.

Procedure 16.3

Test the adaptive advantages of four feeding morphologies.

1. Divide into groups of four students each. Each of the four students must have a different feeding tool.
2. Obtain one food supply for your group consisting of a small container filled with food items.
3. Examine the size of the food item (*Food item* A). Hypothesize which of the available tools is best adapted to gather the food available.
4. All four organisms (group members) will "feed" from the same container placed in the middle of the table equidistant from each organism. A feeding session will last 20 seconds. All organisms will feed at the same time from the same food container.
5. Obtain four small cups, one for each organism. Each organism will feed into a "stomach" represented by the cup kept directly in front of the organism and at the outer edge of the table at all times.
6. Feed for one 20-sec session (*Feeding session 1*).

Table 16.1

Adaptations of External Features of Members of the Major Classes of Vertebrates

	Specimens				
	Fish	Amphibian	Reptile	Bird	Mammal
Protection					
Sensory					
Locomotion					
Gas exchange					

7. Count the number of food items obtained and record the value in table 16.2 for each organism. Return the gathered food to the central container.
8. Rotate feeding tools among the team members and feed for a second 20-sec session. Record the results in table 16.2.
9. Repeat steps 7–8 until all four organisms have used all four beaks (four sessions). Record the results of each session in table 16.2.
10. Select a food supply with a different size food item (*Food item* B). Repeat steps 6–9.
11. Select a food supply with a different size food item (*Food item* C). Repeat steps 6–9.

Questions 9

Which beak is best adaptive to gather *Food item* A? ______

______

Food item B? ______

Food item C? ______

Would a mixture of food sizes be more realistic of a natural situation? ______

______

Is competition a factor in the success of adaptations? Why or why not? ______

______

Does the success (adaptive advantage) of a beak depend on which organism wields that beak? What is your evidence?

______

______

Would the effectiveness of an adaptation for feeding increase with experience by the organism? How so? ______

______

Would a mixture of food sizes amplify or diminish the difference among success of adaptations? ______

______

Table 16.2

Experimental Data Testing the Effectiveness of Four Varied Adaptations

	Adaptations			
Food Item A	Beak 1	Beak 2	Beak 3	Beak 4
Feeding session 1	_____ food items	_____ food items	_____ food items	_____ food items
Feeding session 2	_____ food items	_____ food items	_____ food items	_____ food items
Feeding session 3	_____ food items	_____ food items	_____ food items	_____ food items
Feeding session 4	_____ food items	_____ food items	_____ food items	_____ food items
	mean items per session _____	mean items per session _____	mean items per session _____	mean items per session _____
Food Item B	Beak 1	Beak 2	Beak 3	Beak 4
Feeding session 1	_____ food items	_____ food items	_____ food items	_____ food items
Feeding session 2	_____ food items	_____ food items	_____ food items	_____ food items
Feeding session 3	_____ food items	_____ food items	_____ food items	_____ food items
Feeding session 4	_____ food items	_____ food items	_____ food items	_____ food items
	mean items per session _____	mean items per session _____	mean items per session _____	mean items per session _____
Food Item C	Beak 1	Beak 2	Beak 3	Beak 4
Feeding session 1	_____ food items	_____ food items	_____ food items	_____ food items
Feeding session 2	_____ food items	_____ food items	_____ food items	_____ food items
Feeding session 3	_____ food items	_____ food items	_____ food items	_____ food items
Feeding session 4	_____ food items	_____ food items	_____ food items	_____ food items
	mean items per session _____	mean items per session _____	mean items per session _____	mean items per session _____

Procedure 16.4

Conduct a self-designed test of adaptation effectiveness.

1. Choose one of the following two questions to answer using the general protocol in Procedure 16.3:

   Does the effectiveness of an adaptation for feeding increase with experience by the organism?

   Does a mixture of food sizes amplify or diminish the difference among success of adaptations?

2. Design and execute a procedure to answer your question.
3. Copy and modify table 16.2 as needed for your procedure.

Question 10

What was the answer to your question in Procedure 16.4?

____

____

Questions for Further Thought and Study

1. What functions other than feeding might the shape of a bird's beak serve?

2. Could every characteristic of an organism be considered an adaptation? How so?

3. What is wrong with the statement "This adaptation evolved to promote reproduction"?

4. What function is the hand of a chimpanzee adapted to perform?

17

Adaptations of Plants to Their Environment

Objectives

As you complete this lab exercise you will:

1. Understand plant adaptations shaped by natural selection.
2. Examine morphological plant characteristics and link their structure with an adaptive function.
3. Discover how one structural feature can confer multiple adaptive advantages.

Plants interact with their environment. As a result, evolution driven by natural selection shapes structures and strategies that maximize a plant's survival and reproduction and therefore shape their ecology (fig. 17.1).

Plant adaptations are characteristics and structures of a species that facilitate survival and fitness by enhancing vital processes such as homeostasis, nutrient and gas acquisition, or reproduction. Although adaptations can be metabolic, the most easily examined are structural features.

The number of adaptations occurring in plants is immense. It can be argued that *most* characteristics of a plant species are, at least in part, adaptations that play a role in promoting a major life process. A diffuse root system may be an adaptation that promotes stability in loose soil. Flowering in the spring may be an adaptation to exploit the increasing pollinator (insect) population and therefore enhance fitness (reproductive success). The list of adaptations is almost endless.

Morphological features are the most obvious adaptations, but any characteristic whose form or function promotes survival and reproduction in response to selective pressures of the environment is an adaptation A unique enzyme that produces a compound toxic to caterpillars is an adaptation to prevent herbivory. Spines of a cactus are adapted to fend off grazers. The rapid closing of a leaf-trap of a Venus Flytrap is an adaptation to catch fast-moving insects. The requirement of fire to open the cone and release seeds of some species of pine is an adaptation that synchronizes seed release with newly available nutrients, more available high-quality light, reduced competition, and open soil, all of which increase the chances of germination and success.

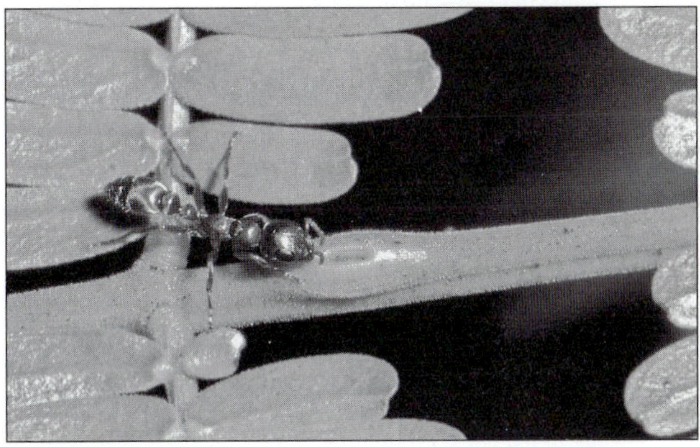

Figure 17.1
Leaves of this acacia plant in Belize have numerous nectaries that produce a sugary liquid. These nectaries are adaptive because they attract ants that protect the plant from caterpillars and other leaf eaters.

In this lab exercise you will examine a range of plant adaptations for gas exchange, water relations, and light acquisition.

GAS EXCHANGE

All organisms exchange gas with their environment. Autotrophic plants depend on CO_2 as a carbon source and O_2 as an electron acceptor for respiration. They also release O_2 as a by-product of photosynthesis.

Procedure 17.1

Examine adaptations for gas exchange.

1. Examine a prepared slide of a privet (*Ligustrum*) leaf cross section that shows common structural adaptations for gas exchange. Locate the stomates (pores) that allow gas exchange at the leaf surface.

Questions 1

Are the pores most abundant on the upper or on the lower surface of the leaf? ________________________

____

Open stomates are needed for gas exchange, but can also allow loss of water vapor. On which surface would it be more adaptive for the stomates to occur to minimize water loss? Why? ____________________

____

Notice that the leaf interior is not a solid mass of cells. What percentage of the cross-sectional area is open space for gas movement? Be sure to examine three or four prepared slides to provide an accurate estimate.____________________

____

2. Examine a prepared slide of a leaf cross section from the water lily *Nymphaea* (fig. 17.2). Aquatic plants are well adapted with "air" pockets within their tissues to supplement gas exchange because the concentration of O_2 in air is 25,000× greater than O_2 dissolved in water.

Questions 2

Are air pockets evident in the leaf cross section of water lily? ____________________

____

What percentage of the cross-sectional area of a water lily leaf includes air spaces? Is this percentage greater than that for the terrestrial privet leaf? ____________________

____

Are stomates apparent in water lily leaves? ____________________

____

Would you expect much direct recycling of gases between respiration and photosynthesis within the leaf of a water lily? How so? ____________________

____

3. Examine a prepared slide of an elderberry stem (*Sambacus*) lenticels.

Questions 3

How does a lenticel appear adaptive for gas exchange?

____

____

Are the cortex cells just inside the lenticels loose with small air spaces? ____________________

____

Not all stems have lenticels. Do they still require gas exchange? Through what path? ____________________

____

WATER RELATIONS

Water availability, more than any other environmental factor, governs the distribution and abundance of plants.

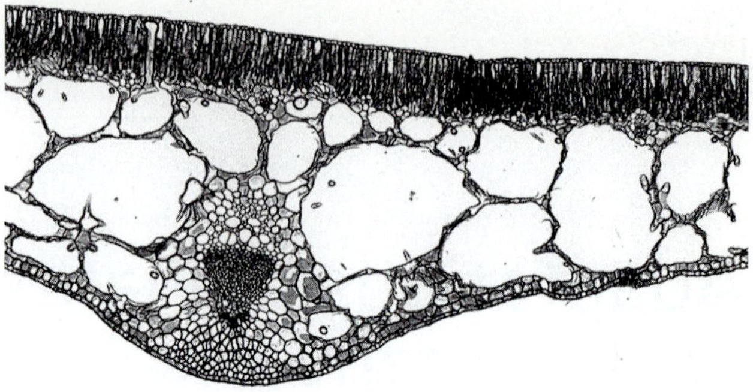

Figure 17.2

This cross section of a water lily leaf reveals extensive, gas-filled chambers. These atmospheric chambers are adaptive because they increase availability of oxygen. These pockets of air hold a higher concentration of oxygen than will dissolve in water.

Hydrophytes are plants adapted to aquatic environments. **Mesophytes** are terrestrial plants adapted to moderate water availability. **Xerophytes** are adapted to low water availability. Terrestrial mesophytes and xerophytes must be well-adapted to acquire, transport, and conserve water.

Procedure 17.2

Examine adaptations for water relations.

1. Examine a prepared slide of a root cross section from the mesophyte buttercup (*Ranunculus*). The star-shaped cluster of cells in the center of a buttercup root is xylem cells.

Questions 4

How big are the xylem cells of buttercup root relative to the other cells? ____________________

____

Lignin, a reinforcing molecule in the cell walls, typically stains red. Are the xylem cells of buttercup root reinforced?

____

____

How are the cell walls and size of the xylem cells adapted to transport water? ____________________

____

Most functioning xylem cells are hollow. Do the xylem cells of buttercup root appear empty? ____________________

____

If water can diffuse from one cell to another, what is the advantage of having hollow conducting cells? ____________________

____

2. Examine a prepared slide of a corn (*Zea*) root and stem cross sections.

Questions 5

What adaptive characteristics distinguish the xylem cells of corn? ______________________________

____

Vascular bundles are scattered across a corn stem. How were they arranged in the root? ______________________________

____

What characteristic readily distinguishes the water-conducting xylem cells in each vascular bundle? ________

____

Are the vascular bundles rich with structurally reinforcing lignin? How is that adaptive? ______________________________

____

3. Examine a prepared slide of a stem cross section of the hydrophyte *Elodea*.

Question 6

How does the vascular bundle of *Elodea* compare to that of a mesophyte? How would this be adaptive for *Elodea*?

____

____

4. Wilting leaves is an adaptive response to low water. Wilting constricts the space into which evaporation from stomates occurs. Examine two sunflower plants—one well-watered and the other dry and wilted.

Questions 7

When most dicots wilt, they droop and thereby enclose the lower surface. What is the adaptive significance of enclosing the lower surface? ______________________________

____

Drooping leaves of a mesophyte also temporarily crush or kink the xylem. How is this adaptive? ______________________________

____

5. Most grasses are xerophytes. Examine some well-watered grass leaves and some wilted grass (fig. 17.3).

Question 8

Do wilted grass leaves droop? Or do they tend to curl and enclose a surface? ______________________________

____

6. Examine a prepared slide of a leaf cross section of *Poa*, a common grass. Locate the large buliform cells on either side of the midvein and the stomates on the surface.

Questions 9

On which surface do the stomates occur? ______________________________

____

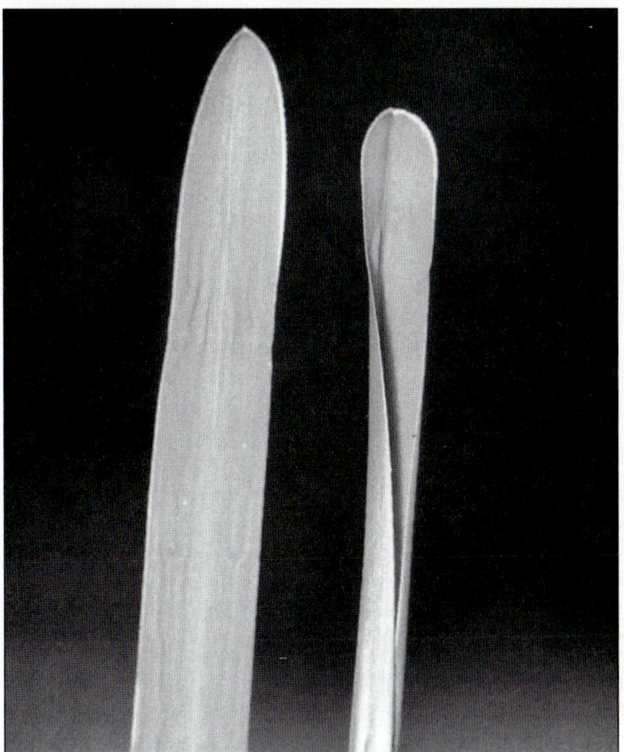

Figure 17.3

The grass leaf on the left has received plenty of water to retain its shape to capture sunlight. The leaf on the right has wilted. Notice that the wilted leaf curls rather than droops. The curl encloses surface with the most stomates.

If the buliform cells lose their turgor pressure and shrink, how would the leaf shape change? ______________________________

____

Does the orientation of the curling of grass leaves appear to crush the xylem as occurred in the sunflower leaves? _____

____

One adaptive response to low water is for the leaves to droop, enclose the stomates, and effectively stop all water flow through the xylem. A contrasting strategy is to curl and enclose the stomates, but allow water to flow down to the last drop. Which appears to be the common mesophyte strategy? ______________________________

____

Which is the common xerophyte strategy? Is one strategy the "correct" one? ______________________________

____

7. Succulents are xerophytes adapted for water storage. Examine a prickly pear cactus (*Opuntia*) stem cross section. In *Opuntia*, the stem grows in "pads" that function as leaves.

Questions 10

Are stomates abundant in *Opuntia*? ________________

____

Are the inner cells large and thin walled, or small and thick walled? ________________

____

8. Most succulent xerophytes are adapted to hold their shape and not wilt. Examine a prepared slide of a leaf cross section of *Yucca*, a succulent xerophyte.

Question 11

Rigid, supportive fibers stain dark red. Do you suspect that leaves of *Yucca* resist wilting? ________________

____

9. In privet leaves, wilting and drooping encloses the air outside of the stomates. Some leaves also have grooved or in-folded surfaces. Examine a slide of an oleander (*Nerium*) leaf cross section and a yucca (*Yucca*) leaf cross section.

Question 12

Where are the stomates located on Oleander and Yucca leaves? How is this adaptive? ________________

____

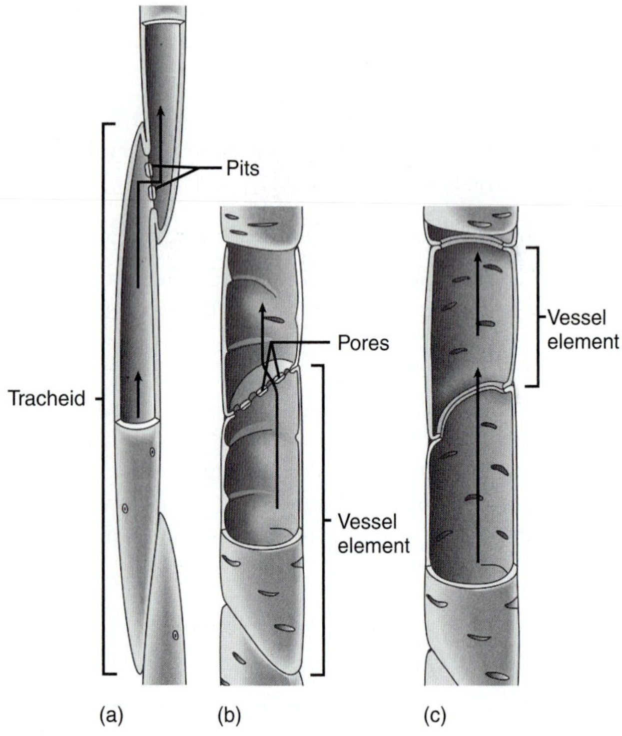

Figure 17.4

Comparison of tracheids and vessel elements. (a) In tracheids, water passes from cell to cell through pits. (b, c) In vessel elements, water moves through pores, which may be simple or interrupted by bars.

10. The xylem of cone-bearing gymnosperms includes elongated cells called *tracheids* through which water passes from cell to cell through pores (fig. 17.4). The evolution of angiosperms included adaptive vessel elements with open ends that form long tubes. Examine a prepared slide of a stem cross section of pine (*Pinus*), a gymnosperm, and of oak (*Quercus*), an angiosperm.

Questions 13

Wood is primarily older, lignin-filled xylem cells that no longer conduct water. Are the alternating rings of large-celled spring wood and dense, fibrous summerwood apparent? ________________

____

During which season is the most growth and water transport likely occurring for pine and oak? ________________

____

During which season (spring or summer) would the largest xylem be produced? ________________

____

Which species has large, open-vessel elements? ________

____

What is the relative cross-sectional area of vessels versus tracheids? How is this adaptive? ________________

____

How are open-ended vessels forming long tubes more adaptive than porous tracheids? ________________

____

11. Line the bottom of a petri plate with a thoroughly wet, doubled paper towel. Sprinkle a dozen lettuce seeds onto the towel. Cover the plate and place it in the dark for 48 h, or until the seeds have germinated. Use a stereoscope to examine the root hairs near the tip of the root (fig. 17.5).

Questions 14

What is the adaptive advantage of having root hairs?

____

____

Each root hair is an extension of a single epidermal cell. How long are the longest of the root hairs? Use a clear ruler marked in millimeters for comparison. ________________

____

Estimate the number of root hairs on a root tip. ________

____

Figure 17.5
Germinated lettuce seed with root hairs.

12. Root surface area is important for water absorption. The surface area of a root 1 cm long with *no* root hairs is about 0.3 cm^2. Dense root hairs will increase surface area of a root as much as 100-fold.

Question 15
How long must a root with root hairs be to have the same surface area as a petri plate (10 cm dia.)? ____________

____

13. Examine the general morphology of some freshly picked leaves of live oak (*Quercus*). The stomates occur on the lower surface.

Questions 16
Is the leaf perfectly flat or is it curved at the edges? ______

____

Does the curve enclose the lower or upper surface? ______

____

During a windy day, over which surface of the curved leaf would the air move fastest? Slowest? ________________

____

How is the curved profile of an oak leaf adaptive? ______

____

14. Use a stereoscope to examine both surfaces of an oak leaf. You may also prepare a wet mount for a compound microscope if you need more magnification.

Questions 17
Which surface is glabrous (smooth) and which is pubescent (hairy) with minute trichomes? ________________

____

What is the shape of an oak-leaf trichome? __________

____

What is adaptive significance of the density of these surface trichomes? ______________________________

____

15. Use a stereoscope to examine the upper surface of an oak leaf. A strong backlight shining from underneath and through the leaf will make the network of veins more apparent. Find the patches of photosynthetic cells that appear as islands bordered by vascular tissue.

Questions 18
What is the maximum distance between a photosynthetic cell and a nearby vascular bundle? You may need a small metric ruler in the field of view while you examine the surface. ______________________________

____

What is the adaptive advantage of this short distance?

____

____

LIGHT ACQUISITION

Procedure 17.3

Examine adaptations for light acquisition.

1. Examine a prepared slide of a privet (*Ligustrum*) leaf cross section.

Questions 19
Against which surface are photosynthetic cells most tightly packed? How is that adaptive? ________________

____

How are the axes of the cells oriented? Vertical to the leaf surface? Or horizontal? ______

______

Which orientation would require the light to go through the least cell wall material to penetrate the leaf and reach the most chloroplasts? ______

______

Is this adaptive? ______

______

2. The stems of trees are rarely photosynthetic. Examine a cross section of a pine (*Pinus*) or basswood (*Tilia*) stem.

Questions 20

For what function is a tree stem best adapted? What is your evidence from examining this cross section? ______

______

How might this function also be adaptive for the tree's light gathering? ______

______

3. We might expect a leaf to respond to low light by producing more chlorophyll to capture as much limited light as possible, or by producing less chlorophyll because it is not needed. Examine leaves of a bean plant (*Phaseolus*) exposed to low light and a plant exposed to intense light.

Questions 21

Is the ability to vary chlorophyll synthesis an adaptation? How so? ______

______

Which leaves have synthesized the most chlorophyll? How might this response to low light be adaptive? ______

______

4. Closely examine the stems of bean plants grown in low light.

Question 22

Are stems longer in low light plants? How might this response to low light be adaptive? ______

______

INTEGRATION OF ADAPTATIONS

Many morphological adaptations have an obvious and single functional advantage, but it's not always that simple. Extensive root systems, for example, are obviously adaptive for absorbing water from the soil. But, tall trees gathering light at the top of a forest canopy also benefit from extensive root support. So, is an extensive root system an adaptation for absorbing water or for maximum light gathering? Or both? Indeed, major adaptations are best understood when considering the total success of the whole organism. One adaptation can have multiple advantages.

Procedure 17.4

Examine patterns of integrated adaptations.

1. Briefly describe the integrated roles of air spaces within the leaves of hydrophytes for each of the following processes of aquatic plants:
   Gas exchange ______
   Light acquisition ______
   Water relations ______
2. Briefly describe the integrated roles of thick-walled xylem cells within plant stems for each of the following processes:
   Gas exchange ______
   Light acquisition ______
   Water relations ______
3. Briefly describe the integrated roles of stomates of plant leaves for each of the following processes:
   Gas exchange ______
   Light acquisition ______
   Water relations ______
4. Briefly describe the integrated roles that wilting and enclosure of the stomatal surface play for each of the following processes:
   Gas exchange ______
   Light acquisition ______
   Water relations ______

Questions for Further Thought and Study

1. Are any characteristics adaptive in some situations but maladaptive in others? How so?

2. If stomates are adaptive for gas exchange, why don't plants evolve more and more stomates with each generation?

3. Reproduction is the most vital of all plant processes. What are some common plant adaptations that promote successful reproduction?